Emerging Topics in Pattern Recognition and Artificial Intelligence

Series on Language Processing, Pattern Recognition, and Intelligent Systems

Print ISSN: 2661-4316
Online ISSN: 2661-4324

Co-Editors

Ching Y. Suen
Concordia University, Canada
parmidir@enes.concordia.ca

Lu Qin
The Hong Kong Polytechnic University, Hong Kong
csluqin@comp.polyu.edu.hk

Published

Vol. 9   *Emerging Topics in Pattern Recognition and Artificial Intelligence*
         edited by Mounîm A El-Yacoubi, Nicole Vincent and Camille Kurtz

Vol. 8   *Frontiers in Bioimage Informatics Methodology*
         edited by Jie Zhou, Hanchuan Peng and Marianna Rapsomaniki

Vol. 7   *Image Analysis and Pattern Recognition: State of the Art in the Russian Federation*
         edited by Igor Gurevich and Vera Yashina

Vol. 6   *Advances in Pattern Recognition and Artificial Intelligence*
         edited by Nicola Nobile, Marleah Blom and Ching Y. Suen

Vol. 5   *Frontiers in Pattern Recognition and Artificial Intelligence*
         edited by Marleah Blom, Nicola Nobile and Ching Y. Suen

Vol. 4   *Computational Linguistics, Speech and Image Processing for Arabic Language*
         edited by Neamat El Gayar and Ching Y. Suen

Vol. 3   *Social Media Content Analysis:*
         Natural Language Processing and Beyond
         edited by Kam-Fai Wong, Wei Gao, Wenjie Li and Ruifeng Xu

Vol. 2   *Advances in Chinese Document and Text Processing*
         edited by Cheng-Lin Liu and Yue Lu

Vol. 1   *Digital Fonts and Reading*
         edited by Mary C. Dyson and Ching Y. Suen

Series on Language Processing, Pattern Recognition,
and Intelligent Systems — Vol. 9

Emerging Topics in Pattern Recognition and Artificial Intelligence

Editors

Mounîm A El-Yacoubi
Institut Polytechnique de Paris, France

Nicole Vincent
Université Paris Cité, France

Camille Kurtz
Université Paris Cité, France

NEW JERSEY · LONDON · SINGAPORE · BEIJING · SHANGHAI · HONG KONG · TAIPEI · CHENNAI · TOKYO

Published by

World Scientific Publishing Co. Pte. Ltd.
5 Toh Tuck Link, Singapore 596224
USA office: 27 Warren Street, Suite 401-402, Hackensack, NJ 07601
UK office: 57 Shelton Street, Covent Garden, London WC2H 9HE

Library of Congress Control Number: 2024002426

British Library Cataloguing-in-Publication Data
A catalogue record for this book is available from the British Library.

Series on Language Processing, Pattern Recognition, and Intelligent Systems — Vol. 9
EMERGING TOPICS IN PATTERN RECOGNITION AND ARTIFICIAL INTELLIGENCE

Copyright © 2025 by World Scientific Publishing Co. Pte. Ltd.

All rights reserved. This book, or parts thereof, may not be reproduced in any form or by any means, electronic or mechanical, including photocopying, recording or any information storage and retrieval system now known or to be invented, without written permission from the publisher.

For photocopying of material in this volume, please pay a copying fee through the Copyright Clearance Center, Inc., 222 Rosewood Drive, Danvers, MA 01923, USA. In this case permission to photocopy is not required from the publisher.

ISBN 978-981-12-8911-8 (hardcover)
ISBN 978-981-12-8912-5 (ebook for institutions)
ISBN 978-981-12-8913-2 (ebook for individuals)

For any available supplementary material, please visit
https://www.worldscientific.com/worldscibooks/10.1142/13752#t=suppl

Desk Editors: Balasubramanian Shanmugam/Steven Patt

Typeset by Stallion Press
Email: enquiries@stallionpress.com

© 2025 World Scientific Publishing Company
https://doi.org/10.1142/9789811289125_fmatter

Preface

Contributions in Pattern Recognition and Artificial Intelligence gathers a collection of topics of general interest in the area of pattern recognition and artificial intelligence (AI), particularly advances in machine and deep learning. This book is of interest to students, academics, industrials researchers in the field, as well as the general public.

This volume consists of a collection of chapters written by a diverse range of international scholars. Papers and presentations were carefully selected from 153 papers from the International Conference on Pattern Recognition and Artificial Intelligence (ICPRAI) held in Paris, France (June 2022). The conference and lecture were hosted by the Université Paris Cité. The event was well received, with over 165 attendees from 31 countries.

The chapters of this volume provide readers with an overview of research works highlighting the interlink between pattern recognition and artificial intelligence, with a particular emphasis on machine and deep learning. The selected works share key information on relevant topics of interest, related to pattern recognition and AI, namely mathematic modeling with focus on machine and deep learning, signal and image processing, as well as the application fields. The chapters are grouped into four categories, namely explainable AI, graphs, applications, and segmentation.

The first two chapters address the first category, explainable AI, a recent topic that seeks to make AI models interpretable and to make the prediction of these models understandable to stakeholders and laymen. In Chapter 1, Fuchs and Riesen address the task of graph augmentation using matching-graphs. In Chapter 2, Kropatsch and his coauthors detail their work on controlling topology preserving graph pyramids. This chapter is related to Kropatsch's keynote speech at ICPRAI 2022.

In the second category, Chapters 3, 4, and 5 address several AI tasks based on graphs, a data representation that is well suited to a wide range of real-life data, naturally represented as graphs. In Chapter 3, Alexiadis and his coauthors detail their sensor-independent multimodal approach scheme for human activity recognition. In Chapter 4, Gomez and his coauthors propose metrics for saliency map evaluation of deep learning explanation methods. In Chapter 5, Rojas and his coauthors describe their work on efficient segmentation of e-waste devices with deep learning for robotic recycling.

In the third category of this volume, several applications related to pattern recognition and AI are presented. In Chapter 6, Almuhajri and Suen propose an approach for shop signboards' detection using the ShoS dataset. In Chapter 7, Mendoza and Pedrini present a self-distilled self-supervised scheme for monocular depth estimation while in Chapter 8, Lin and his coauthors propose an encoder-decoder approach for offline handwritten mathematical expression recognition with residual attention. In Chapter 9, Horváth and Kontár present a complexity analysis of the general feasibility of patch-based adversarial attacks on semantic segmentation problems.

In the fourth category, segmentation, Chapters 10, 11, and 12 cover several segmentation-related tasks and applications. In Chapter 10, related to NLP, Bansal and his coauthors propose a sentiment and word cloud analysis of tweets related to COVID-19 vaccines before, during, and after the second wave in India while in Chapter 11, Chopin and his coauthors propose reinforcement learning and sequential QAP-based graph matching for semantic segmentation of images. In Chapter 12, Zhukov and his coauthors propose feature explanation methods with statistical filtering of important features.

The publication of this book has been possible thanks to the valuable contributions of the authors. We would like also to thank the reviewers for their timely and relevant reviews that helped enhance the quality of the chapters.

About the Editors

Camille Kurtz obtained a PhD degree in computer science in 2012 from the Université de Strasbourg (France). After a postdoctoral period at Stanford University (USA), he was appointed Assistant Professor at Université Paris Descartes (France). Since 2023, he is now Full Professor at Université Paris Cité (France) in the computer science department. His research interests are centered on image representations, with visual, semantic, and perceptual aspects, with applications in medical imaging, remote sensing, and image/video analysis.

Nicole Vincent, joining mathematics and computer sciences studies, completed his PhD in 1988 and became Full Professor in 1996. She was Vice-head of her lab in 1998 and Head of her lab in 2004. She participated in many local and more general boards in her university and was a member of the board of governors of Paris Descartes University from 2016 to 2019. She has been involved in several conference organizations and has been a member of the editorial board of the *IJPRAI* journal as well as *Pattern Recognition* journal. Her research is organized into three main research topics, in the realm of computer vision and image analysis, and more precisely in the analysis of documents, in particular handwriting, quality of documents, old documents, word spotting, and security, in medical image analysis, in particular ultrasound images and mammograms, and in video analysis, in particular tracking and text extraction.

Mounîm A. El-Yacoubi is a Professor at Institut Polytechnique de Paris and Institut Mines Telecom since 2008. Prior to that, he has done research and developed, at the French Post Research Center, Centre for Pattern Recognition and Machine Intelligence (Concordia Univ., Canada), and Parascript (USA), handwriting recognition software for mail sorting, bank check reading, and form processing, still running daily in automatic reading machines worldwide. He was Program Chair of ICPRAI 2022, IEEE SWC 2023, and ICCSI 2023 and is General Chair of the 16th International Conference on Human System Interaction (HSI), 2024. His research interests include AI, machine and deep learning, pattern recognition, and modeling human data, especially behavioral data like handwriting, voice, and gestures. Applications include biometrics, sports analytics, human mobility analysis, and e-health, in particular the detection of neurodegenerative and chronic diseases, like Parkinson's, Alzheimer's, and diabetes.

Contents

Preface	v
About the Editors	ix

Chapter 1. Graph Augmentation Using Matching-Graphs 1
Mathias Fuchs and Kaspar Riesen

Chapter 2. Controlling Topology Preserving Graph Pyramids 27
Walter G. Kropatsch, Majid Banaeyan, and Rocio Gonzalez-Diaz

Chapter 3. A Sensor-Independent Multi-Modal Fusion Scheme for Human Activity Recognition 77
Anastasios Alexiadis, Alexandros Nizamis, Dimitra Zotou, Dimitrios Giakoumis, Konstantinos Votis, and Dimitrios Tzovaras

Chapter 4. Metrics for Saliency Map Evaluation of Deep Learning Explanation Methods 101
Tristan Gomez, Thomas Fréour, and Harold Mouchère

Chapter 5. Efficient Segmentation of E-Waste Devices With Deep Learning for Robotic Recycling   123

Cristof Rojas, Antonio Rodríguez-Sánchez, and Erwan Renaudo

Chapter 6. Shop Signboard Detection Using the ShoS Dataset   145

Mrouj Almuhajri and Ching Y. Suen

Chapter 7. Self-Distilled Self-Supervised Monocular Depth Estimation   165

Julio Mendoza and Helio Pedrini

Chapter 8. An Encoder–Decoder Approach to Offline Handwritten Mathematical Expression Recognition with Residual Attention   187

Qiqiang Lin, Chunyi Wang, Ning Bi, Ching Y Suen and Jun Tan

Chapter 9. A Complexity Analysis on the General Feasibility of Patch-Based Adversarial Attacks on Semantic Segmentation Problems   205

András Horváth and Soma Kontár

Chapter 10. Sentiment and Word Cloud Analysis of Tweets Related to COVID-19 Vaccines before, during, and after the Second Wave in India   229

Anmol Bansal, Arjun Choudhry, Anubhav Sharma, and Seba Susan

Chapter 11. Reinforcement Learning and
Sequential QAP-Based Graph Matching
for Semantic Segmentation of Images   259

*Jérémy Chopin, Jean-Baptiste Fasquel,
Harold Mouchère, Rozenn Dahyot and Isabelle Bloch*

Chapter 12. FEM and Multi-Layered FEM:
Feature Explanation Methods with
Statistical Filtering of Important Features   295

*Alexey Zhukov, Jenny Benois-Pineau, Romain Giot,
Romain Bourqui, and Luca Bourroux*

© 2025 World Scientific Publishing Company
https://doi.org/10.1142/9789811289125_0001

Chapter 1

Graph Augmentation Using Matching-Graphs

Mathias Fuchs* and Kaspar Riesen

Institute of Computer Science, University of Bern, Switzerland
*mathias.fuchs@unibe.ch

Abstract

We are observing rapid developments in many areas of intelligent information processing. The reasons for this are that both data access and data acquisition have become simpler and more convenient over the last years. In many cases, the underlying data are complex, making vectorial structures rather unsuitable for data representation. In these cases, graphs offer a versatile alternative to purely numerical approaches. Regardless of the actual representation formalism used, it is inevitable for supervised pattern recognition algorithms to have access to large sets of labeled training patterns. In some cases, however, this requirement cannot be met because the set of labeled samples is inherently limited. In a current research project, a novel encoding of pairwise graph matchings is introduced. The basic idea of this encoding is to formalize the stable cores of pairs of graphs using so-called matching-graphs. In this chapter, we propose a new scenario for using these matching-graphs. In particular, we use them to augment training sets of graphs to make the training of a classifier more robust. The benefit of this approach is empirically validated in two different experiments. First, we study the augmentation approach on very small graph datasets in conjunction with an SVM classifier, and second, we study the augmentation approach on data sets with reasonable size in conjunction with a graph neural network classifier.

1.1 Introduction and Related Work

Pattern recognition is a scientific discipline that aims to algorithmically solve various problems such as emotion recognition [1], person re-identification [2], or signature verification [3], to name just three prominent examples. The very first step in any pattern recognition scenario is to solve the data representation challenge. That is, one has to decide how to represent the underlying entities so that they can be processed automatically by machines. In terms of data representation, graphs offer a versatile alternative to feature vectors because they can encode both entities and their relationships at the same time. Actually, due to their power and flexibility, graphs have found widespread application in pattern recognition and related areas (these applications range from gait recognition [4], over link-fault detection [5], to object recognition [6]).

A substantial part of the graph-based pattern recognition methods available is based on some sort of *graph matching* [7]. The overall aim of graph matching is to find a correspondence between the nodes and edges of two graphs that satisfies some, more or less, stringent constraints. *Graph edit distance* [8, 9] is acknowledged as one of the most flexible graph-matching models available to date. The major advantage of this paradigm over other distance measures (like *graph kernels* [10] or *graph neural networks* [11]) is that graph edit distance provides more information than merely a pairwise dissimilarity or similarity score. In particular, graph edit distance indicates which sub-parts of the underlying graphs actually correspond to each other (known as *edit path*). In a recent work [12], the authors of this chapter suggest to explicitly use this matching information to encode stable parts of pairs of graphs in a novel data structure called *matching-graph*.

In 1985, Robert Mercer made his famous comment "There is no data like more data" [13]. Some researchers even go one step further by arguing that having more data is more important than developing better algorithms [14]. At least one can agree that labeled training data is one of the most pivotal prerequisites for the development and evaluation of supervised pattern recognition and machine learning algorithms. This is particularly true for deep learning methods [15, 16], which typically perform better the more examples of a given phenomenon a network is exposed to. However, in real-world applications, we are often faced with the fact that the amount of labeled training data is limited (for various reasons, but probably most frequently for cost and/or time reasons, which are consumed in the manual classification of data).

The main contribution of this chapter is that we introduce and research a novel and systematic way of increasing the amount of training data in graph-based pattern recognition scenarios. The basic idea is to compute matching-graphs for any pair of training graphs available. By additionally slightly randomizing the whole process of creating the matching-graphs, the amount of training data can hereby be increased to virtually any size. Our main hypothesis is that this novel method provides a natural way to create realistic and relevant graphs that are actually useful during training of pattern recognition algorithms (note that this process is evaluated for relatively small graphs with a low edge density in binary classification scenarios).

The proposed process of creating matching-graphs in order to enlarge training sets is similar in spirit to existing graph augmentation approaches [17, 18]. However, most of these approaches augment the graphs by altering edge information only. Moreover, these approaches often rely on a single sample of a graph. In our approach, however, we generate new graphs based on the information captured in the edit path resulting from matching pairs of graphs. The graphs generated this way include both edge and node modifications. In [19], an adaptation of the *Mixup algorithm* [20] for graphs is proposed, which is quite similar to our method. In addition, there are several neural network-based approaches that are somewhat similar to our proposal but do not have the main goal of improving graph classification. Several approaches attempt, for instance, to leverage the power of *Variational Auto-Encoders* (VAEs) for graph generation [21, 22]. Last but not least, the success of *Generative Adversarial Networks* (GANs) for image generation led to an adaptation for graph generation as well [23, 24].

The remainder of this chapter, which actually combines and extends two preliminary papers [25, 26],[a] is organized as follows.

---

[a]In [25], we describe the novel concept of matching-graphs and use them to augment small graph datasets in conjunction with an SVM classifier. In [26], we use the matching-graphs in order to augment regularly sized graph datasets in conjunction with graph neural networks.

Section 1.2 makes this chapter self-contained by providing basic definitions and terms used throughout this chapter. Next, in Section 1.3, a detailed description of the data augmentation process is given in conjunction with the general procedure of creating matching-graphs. Eventually, in Section 1.4, we provide empirical evidence that our approach of generating training samples is able to improve the classification accuracy of existing classification systems. Finally, in Section 1.5, we conclude this chapter and discuss some ideas for future work.

1.2 Basic Definitions

In this section, we first formally define the concept of a graph and then introduce the two basic graph-based frameworks actually used in this chapter, namely graph edit distance and graph neural networks.

1.2.1 *Graph-based representations*

Different definitions of graphs are being used — what they all have in common is that graphs consist of nodes and edges, both of which can possibly have labels. The following definition allows us to handle arbitrarily structured graphs with unconstrained labeling functions.

Definition 1.1 (Graph). Let L_V and L_E be finite or infinite label sets for nodes and edges, respectively. A *graph* g is a four-tuple $g = (V, E, \mu, \nu)$, where V is the finite set of nodes, $E \subseteq V \times V$ is the set of edges, $\mu : V \to L_V$ is the node labeling function, and $\nu : E \to L_E$ is the edge labeling function.

The labels for both nodes and edges can be given, for instance, by a vector space $L = \mathbb{R}^n$ or a set of symbolic labels $L = \{\alpha, \beta, \gamma, \ldots\}$. Given that the nodes and/or the edges are labeled, the graphs are referred to as *labeled graphs*. *Unlabeled graphs* are obtained as a special case by assigning the same (empty) label $\varnothing$ to all nodes and edges, i.e., $L_V = L_E = \{\varnothing\}$. In some algorithms and applications, it is necessary to include *empty "nodes"* and/or *empty "edges"* (also referred to as *null nodes* and *null edges*). We denote both empty nodes and empty edges by ε.

Edges are given by pairs of nodes (u, v), where $u \in V$ denotes the source node and $v \in V$ the target node of a directed edge. Commonly, the two nodes u and v connected by an edge (u, v) are referred to as *adjacent*. *Directed graphs* directly correspond to the above stated definition of a graph. However, the class of *undirected graphs* can be modeled as well, inserting a reverse edge $(v, u) \in E$ for each edge $(u, v) \in E$ with an identical label, i.e., $\nu(u, v) = \nu(v, u)$. In this case, the direction of an edge can be ignored, since there are always edges in both directions.

1.2.2 Graph edit distance (GED)

Computing the similarity or dissimilarity between pairs of entities is a fundamental process used in many pattern recognition systems. In a vector space, the rich set of similarity measures (e.g., the cosine similarity) or dissimilarity measures (e.g., the Euclidean distance) can be used. In the graph domain, however, this choice is somewhat more difficult, since there are also many measures available but none is considered a true gold standard. Nevertheless, one may say that the *graph edit distance* is a widely accepted concept to calculate a flexible graph dissimilarity between pairs of graphs. Actually, in the experimental evaluation of this chapter, we use this particular dissimilarity model in conjunction with a support vector machine for graph classification.

Given two graphs $g = (V, E, \mu, \nu)$ and $g' = (V', E', \mu', \nu')$, the basic idea of graph edit distance is to transform g into g' using some *edit operations*. A standard set of edit operations is given by *substitutions*, *deletions*, and *insertions*, of both nodes and edges. We denote the substitution of two nodes $u \in V$ and $v \in V'$ by $(u \to v)$, the deletion of node $u \in V$ by $(u \to \varepsilon)$, and the insertion of node $v \in V'$ by $(\varepsilon \to v)$. An equivalent notation is used for the three operations on the edges.

A set $\{e_1, \ldots, e_s\}$ of s edit operations e_i that transform a source graph g completely into a target graph g' is called an *edit path* $\lambda(g, g')$ between g and g'. Typically, one assigns a cost of $c(e_i)$ to each edit operation e_i. The idea of such a cost is to define whether or not an edit operation e_i represents a strong modification of the graph. Clearly, between two similar graphs, there should exist an inexpensive edit path, representing low cost operations, while for dissimilar graphs, an edit path with high cost is needed. Consequently, the edit distance of two graphs is defined as follows.

Definition 1.2 (Graph Edit Distance). Let $g_1 = (V_1, E_1, \mu_1, \nu_1)$ be the source and $g_2 = (V_2, E_2, \mu_2, \nu_2)$ the target graph. The *graph edit distance* $d_{\lambda_{\min}}(g_1, g_2)$, or $d_{\lambda_{\min}}$ for short, between g_1 and g_2 is defined by

$$d_{\lambda_{\min}}(g_1, g_2) = \min_{\lambda \in \Upsilon(g_1, g_2)} \sum_{e_i \in \lambda} c(e_i), \tag{1.1}$$

where $\Upsilon(g_1, g_2)$ denotes the set of all complete edit paths transforming g_1 into g_2, c denotes the cost function measuring the strength $c(e_i)$ of edit operation e_i, and $\lambda_{\min}$ refers to the minimal cost edit path found in $\Upsilon(g_1, g_2)$.

Optimal algorithms for computing the edit distance of two graphs are typically based on combinatorial search procedures with exponential time complexity. Thus, applying graph edit distance to large graphs is computationally demanding (or not possible).

In order to reduce the computational complexity of graph edit distance, several approximation algorithms have been proposed in the literature [27, 28]. In this chapter, we use the often employed approximation algorithm BP [29]. This specific algorithm reduces the problem of graph edit distance computation to an instance of a linear sum assignment problem for which several efficient algorithms exist. The approximated graph edit distance between g and g' computed by algorithm BP is termed $d_{\text{BP}}(g, g')$ from now on.

1.2.3 *Graph neural networks (GNNs)*

The second part of the empirical evaluation presented in this chapter is based on the concept of *Graph Neural Networks* [30]. GNNs actually provide a general framework to use deep learning on graph-based data. Roughly speaking, GNNs aim to learn feature vectors h_v or h_g to represent each node $v \in V$ or the complete graph g, respectively. That is, by means of GNNs one obtains an embedding of either single nodes or complete graphs in a vector space. Based on this embedding, the individual nodes or the entire graph can then be classified. In our work, we focus on graph classification only.

In order to learn the graph embedding, we first need to learn the embedding of each node. In order to achieve this goal, GNNs usually follow a neighborhood aggregation strategy. That is, GNNs use a form of *neural message passing* in which messages are exchanged between the nodes of a graph [31]. The general idea is to iteratively update the representation of a node by aggregating the representation of its neighbors. During the i-th message-passing iteration, a hidden embedding $h_v^{(i)}$ that corresponds to each node $v \in V$ is updated according to the aggregated information from its neighborhood. Formally, this can be achieved by two differentiable functions.

- $m_{\mathcal{N}(v)}^{(i)} = \text{AGGREGATE}^{(i)}(\{h_u^{(i)} : u \in \mathcal{N}(v)\},$
- $h_v^{(i+1)} = \text{UPDATE}^{(i)}(h_v^{(i)}, m_{\mathcal{N}(v)}^{(i)}),$

where $m_{\mathcal{N}(v)}^{(i)}$ is the "message" that is aggregated from the neighborhood $\mathcal{N}(v)$ of node v ($\mathcal{N}(v)$ refers to the set of nodes adjacent to v). The feature vector representation of node v at the i-th iteration is $h_v^{(i)}$, and the initial representation $h_v^{(0)}$ defined to be $\mu(v)$, i.e., the original label of node v.

After N iterations of the message passing mechanism, this process produces node embeddings $h_v^{(N)}$ for each node $v \in V$. To get the embedding h_g for the complete graph, so-called *graph pooling* is necessary. Graph pooling combines the individual local node embeddings to one global embedding. That is, the pooling function maps the set of n node embeddings $\{h_{v_1}^{(N)}, \ldots, h_{v_n}^{(N)}\}$ to the graph embedding h_g. The pooling function can simply be a sum (or mean) of the node embeddings or more sophisticated functions [32].

In the context of this chapter, we employ three different GNN architectures that are all based on the above-mentioned concepts. The first architecture, denoted as $Simple_F$ from now on, contains three graph convolutional layers [33]. The second system is the graph isomorphism network, denoted as GIN_F, introduced by Xu et al. [34]. The third architecture is the $GraphSAGE_F$ network introduced by Hamilton et al. [35]. All three algorithms are implemented using

Pytorch Geometric, and for GIN_F and $GraphSAGE_F$, we use the implementations of [36].[b] For the final graph classification, we add a dropout layer to all three architectures and feed the graph embedding into a fully connected layer.

1.3 Augment Training Sets by Means of Matching-Graphs

The major contribution of this chapter is that we propose an approach to increase the size of a given training set. The motivation for this increase comes from the fact that with more and diverse training data, one can usually train more robust classifiers. The proposed method for training set augmentation is based on *matching-graphs*.

Matching-graphs are built by extracting information on the matching of pairs of graphs and by formalizing and encoding this information in a data structure. Matching-graphs can be interpreted as denoised core structures of the underlying graphs. The idea of matching-graphs initially emerged in [12] where they are employed for improving the overall quality of graph edit distance. The matching-graphs used in this chapter are adapted for graph set augmentation and are created in a slightly different way as originally proposed.

Formally, we assume k sets of training graphs $G_{\omega_1}, \ldots, G_{\omega_k}$ stemming from k different classes $\omega_1, \ldots, \omega_k$. For all pairs of graphs stemming from the same class ω_l, the graph edit distance is computed by means of algorithm BP [29]. Hence, we obtain a (sub-optimal) edit path $\lambda(g, g') = \{e_1, \ldots, e_s\}$ for each pair of graphs $g, g' \in G_{\omega_l} \times G_{\omega_l}$. Each edit operation $e_i \in \lambda(g, g')$ can either be a substitution, a deletion, or an insertion of a node including the corresponding edge edit operation.

Based on the edit path $\lambda(g, g')$, two matching-graphs $m_{g \times g'}$ and $m_{g' \times g}$ can now be built (one based on the source graph g and one based on the target graph g'). In order to create both $m_{g \times g'}$ and $m_{g' \times g}$, we initially define $m_{g \times g'} = g$ and $m_{g' \times g} = g'$. Then, we select a certain percentage $p \in [0, 1]$ of all s edit operations available in

---

[b]https://github.com/diningphil/gnn-comparison.

$\lambda(g, g')$. Hence, we obtain a partial edit path $\tau(g, g') = \{e_1, \ldots, e_t\} \subseteq \lambda(g, g')$ with $t = \lfloor p \cdot s \rfloor$ edit operations only. Next, each edit operation $e_i \in \tau(g, g')$ is applied on graphs $m_{g \times g'}$ or $m_{g' \times g}$ according to the following rules[c]:

- If e_i refers to a substitution $(u \to v)$, it is applied on both graphs $m_{g \times g'}$ and $m_{g' \times g}$. More precisely, the labels of the matching nodes $u \in V$ and $v \in V'$ are exchanged in both $m_{g \times g'}$ and $m_{g' \times g}$. Note that this operation shows no effect when the two labels of the involved nodes are identical.
- If e_i refers to a deletion $(u \to \varepsilon)$, e_i is applied on $m_{g \times g'}$ only, meaning that $u \in V$ is deleted in $m_{g \times g'}$.
- If e_i refers to an insertion $(\varepsilon \to v)$, e_i is applied on $m_{g' \times g}$ only. This means that the node $v \in V'$ that is inserted according to the edit operation is deleted in $m_{g' \times g}$ instead.

The rationale for the third rule is as follows. When inserting a node $v \in V'$, it is not necessarily clear how v should be connected to the remaining parts of the current graph (since it is possible that not all necessary edit operations have been carried out or have not been selected at all). By deleting the node $v \in V'$ in $m_{g' \times g}$ rather than inserting it in $m_{g \times g'}$, we can avoid this problem quite easily.

Both matching-graphs represent intermediate graphs between the two underlying training graphs. If p is set to 1.0, all edit operations from the complete edit path $\lambda(g, g')$ are considered during the matching-graph creation. Note, however, that according to our rules, deletions and insertions are uniquely applied on the source or the target graph, respectively. Hence, in this particular parameter setting, we obtain two matching-graphs that are sub-graphs from the original graphs. With parameter values $p < 1.0$, however, we obtain matching-graphs in which possibly some of the nodes are either deleted from g or g' and some other nodes are potentially altered according to their labeling (due to the applied substitutions).

Note that one can extract several partial edit paths $\tau(g, g')$ from one edit path $\lambda(g, g')$ using different values of p. This in turn results in several matching-graphs based on the same edit path. In theory, one

---

[c]Note that these rules can also be applied to edges, as edge operations implicitly follow the node operations. Furthermore, in our case, we use unlabeled edges only.

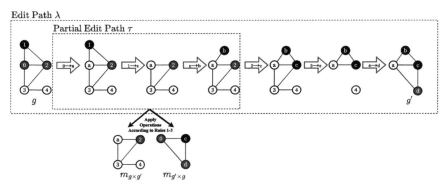

Fig. 1.1. An example of a complete edit path λ, a partial edit path τ, and the resulting matching-graphs $m_{g \times g'}$ and $m_{g' \times g}$.

can create as many matching-graphs as edit operations are available in $\lambda(g, g')$.

Example 1. Figure 1.1 shows a visual example of the graph edit distance between two graphs g and g' and two possible resulting matching-graphs $m_{g \times g'}$ and $m_{g' \times g}$. The corresponding edit path is $\lambda = \{(0 \to a), (1 \to \varepsilon), (\varepsilon \to b), (2 \to c), (3 \to \varepsilon), (4 \to d)\}$. The possible matching-graphs $m_{g \times g'}$ and $m_{g' \times g}$, are created with $p = 0.5$, resulting in the partial edit path $\tau(g, g') = \{(0 \to a), (1 \to \varepsilon), (\varepsilon \to b)\}$, that consists of $t = 3$ edit operations. In this example, it is clearly visible that neither $m_{g \times g'}$ nor $m_{g' \times g}$ is a sub-graph of g or g', respectively.

Note that the proposed process can lead to isolated nodes (as observed, for example, in the second last graph in Figure 1.1). Based on the reasoning that we want to create graphs with nodes that are actually connected, we remove disconnected nodes from the matching-graphs whenever they occur.

Based on the process of creating two matching-graphs for pairs of graphs, we can now define an algorithm to augment a given training set with additional graphs. Algorithm 1 takes k sets of training graphs $G_{\omega_1}, \ldots, G_{\omega_k}$ stemming from k different classes $\omega_1, \ldots, \omega_k$ as input. The two **for** loops accomplish the following. For all pairs of graphs g, g' stemming from the same class ω_i, two matching-graphs $m_{g \times g'}$ and $m_{g' \times g}$ are built and added to the corresponding set of graphs (labeled with G_{ω_i}). Assuming n training graphs per class G_{ω_i}, this results in $k \cdot n(n-1)$ matching-graphs in total, which are directly used to augment the corresponding training sets $G_{\omega_1}, \ldots, G_{\omega_k}$.

In Algorithm 1, the probability $p \in [0,1]$ used for the creation of the matching-graphs (see Line 5) is redefined for each iteration. This can be achieved, for example, by choosing p randomly in each iteration. However, the probability p can also be set to a fixed value for all iterations. In addition to this, inside the second **for** loop, just before the definition of p, a further **for** loop could be defined, such that even more than one matching-graph could be created for each pair of graphs.

Algorithm 1: Graph Augmentation Algorithm
---

input : sets of graphs from k different classes $\mathcal{G} = \{G_{\omega_1}, \ldots, G_{\omega_k}\}$
output: same sets augmented by matching-graphs

1 **foreach** set $G_{\omega_i} \in \mathcal{G}$ **do**
2   M = {}
3   **foreach** pair of graphs $g, g' \in G_{\omega_i} \times G_{\omega_i}$ **do**
4   Compute $\lambda(g, g') = \{e_1, \ldots, e_s\}$
5   Define p in $[0, 1]$
6   Define τ by selecting $\lfloor p \cdot s \rfloor$ edit operations from λ
7   Build both matching-graphs $m_{g \times g'}$ and $m_{g' \times g}$ according to τ
8         $M = M \cup \{m_{g \times g'}, m_{g' \times g}\}$
9   **end**
10     $G_{\omega_i} = G_{\omega_i} \cup M$
11 **end**

1.4 Experimental Evaluation

The overall aim of the following experimental evaluation is to answer the question of whether or not matching-graphs can be beneficially employed as a technique for automatically augmenting training sets of graphs. To this end, we perform two separate experimental evaluations (both applied on the same sets of graphs described in Section 1.4.1). First, in Section 1.4.2, we use the proposed technique to augment very small training sets of graphs and verify whether this helps substantially increase the accuracy of a classification algorithm. Second, in Section 1.4.3, we research whether or not matching-graphs can be used to make the training of graph neural networks more robust.

1.4.1 Datasets

We carry out the experimental evaluation on four different datasets stemming from the repository *Benchmark Data Sets for Graph Kernels*[d] [37]. All data sets employed contain graphs that represent molecules stemming from two classes. By representing atoms as nodes and bonds as edges, graphs can actually represent chemical compounds in a lossless and straightforward manner:

- The **Mutagenicity** dataset is split into two classes, containing mutagenic and non-mutagenic compounds, respectively. Mutagenicity refers to the ability of a chemical compound to cause DNA mutations.
- The **NCI1** dataset [38] originates from anti-cancer screens and is split into molecules that have activity in inhibitioning the growth of non-small cell lung cancer and those that have no activity.
- The third dataset **COX-2** originates from [39] and contains cyclooxygenase-2 (COX-2) inhibitors with or without *in-vitro* activities against human recombinant enzymes.
- The fourth and last dataset **PTC(MR)** stems from the predictive toxicology challenge [40] and consists of compounds that are potentially carcinogenic.

The nodes of all data sets represent the atoms and are labeled with their chemical symbol. The edges of the graphs of the PTC(MR) dataset are labeled with the information about the chemical bonds between the atoms, while the edges of all other sets are unlabeled.

In Table 1.1, we show the total amount of graphs in the dataset, as well as the number of classes, the average number of nodes, and

Table 1.1. The total number of graphs and classes per dataset, as well as the average number of nodes and edges per graph.

Dataset	Graphs	Classes	$\emptyset|V|$	$\emptyset|E|$
Mutagenicity	4,337	2	30.32	30.77
NCI1	4,110	2	29.87	32.30
COX-2	467	2	41.22	43.45
PTC(MR)	344	2	14.29	14.69

---

[d]https://ls11-www.cs.tu-dortmund.de/staff/morris/graphkerneldatasets.

the average number of edges per graph. In our case, the datasets always consist of two classes and the graphs are relatively small with a low edge density.[e]

1.4.2 Augment small training sets

For the first experiment, we artificially decrease the size of all datasets. This is achieved by randomly selecting 10 training graphs per class for each data set. In order to avoid overly simple or very difficult datasets (that might be created by random chance), we repeat the random process of creating small datasets 20 times. The same accounts for training set augmentation by means of our matching-graphs which is also repeated 20 times.

As basic classification system, a Support Vector Machine (SVM) that exclusively operates on a similarity kernel $\kappa(g, g') = -d_{\text{BP}}(g, g')$ is used [41]. Note that any other data-driven classifier could be used in our evaluation as well. However, we feel that the SVM is particularly suitable for our evaluation because of its pure and direct use of the underlying distance information.

The primary reference system is trained on the reduced training data, denoted as $\text{SVM}_R(-d_{\text{BP}})$. Our novel approach, denoted as $\text{SVM}_R+(-d_{\text{BP}})$, is trained on the same training samples of the reduced sets but uses also the created matching-graphs. For the sake of completeness, we also compare our novel framework with a secondary reference system, viz. an SVM that has access to the full training sets of graphs (before the artificial reduction is carried out), denoted as $\text{SVM}_F(-d_{\text{BP}})$.

1.4.2.1 Validation of Metaparameters

In Table 1.2, an overview of all parameters that are optimized is presented. For algorithm BP, which approximates the graph edit distance, the cost for node and edge deletions as well as a weighting parameter $\gamma \in [0, 1]$ that is used to trade-off the relative importance of node and edge edit costs are often optimized [29, 42]. However, for the sake of simplicity, we employ a unit cost of 1.0 for deletions

---

[e]Note that in theory our approach can be applied to any type of graph and dataset where the GED approximation can be calculated in reasonable time.

Table 1.2. Description of all parameters and the corresponding evaluated values.

Parameter	Description	Evaluated Values
γ	Scales node and edge costs for graph edit distance	$\{0.05, 0.10, \ldots, 0.95\}$
p	Relative amount of edit operations selected λ	$\{0.25, 0.50, 0.75, 1.0\}$
C	Weighting parameter of the SVM	$\{10^{-4}, 5 \cdot 10^{-4}, 10^{-3}, 5 \cdot 10^{-3}, 10^{-2}, 5 \cdot 10^{-2}, 10^{-1}, 5 \cdot 10^{-1}, 10^{0}, 10^{1}, 10^{2}\}$

and insertions of both nodes and edges and optimize the weighting parameter γ only (on all datasets).

For the creation of the matching-graphs — actually also dependant on the cost model — the same weighting parameter is independently optimized. Additionally, we optimize the probability p (Algorithm 1, Line 5) of the edit path operations that are used for the matching-graph creation (the optimal value p is then fixed for all iterations). For the SVM classifier itself, parameter C is optimized to trade off between the size of the margin and the number of misclassified training examples.

1.4.2.2 Test Results and Discussion

In Figure 1.2, we show the reference accuracies of $SVM_R(-d_{BP})$ as well as the accuracies of the proposed system $SVM_R+(-d_{BP})$ as bar charts for all 20 random iterations on all four datasets. The iterations are ordered from the worst to the best performing reference accuracy.

On the NCI1 and COX-2 datasets, we observe that the SVM that relies on the matching-graphs performs better than, or at least equal as, the reference systems in all iterations. On the other two datasets, Mutagenicity and PTC(MR), our system outperforms the reference system in 19 out of 20 iterations. In general, we report substantial improvements over the respective baselines for almost all iterations and datasets.

There is a tendency that the improvement is particularly large in iterations where the reference system performs poorly (most likely due to an unfortunate random selection of training samples which

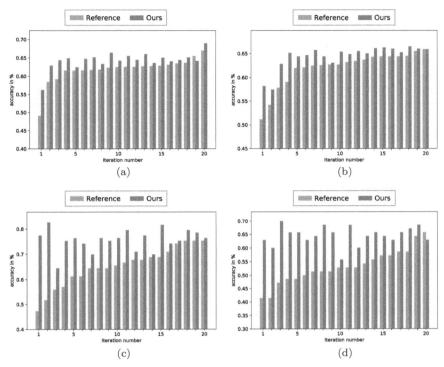

Fig. 1.2. Classification accuracies of all 20 iterations of the reference system (bright bars) compared to our novel system that additionally uses the matching-graphs for training (dark bars). (a) Mutagenicity; (b) NCI1; (c) COX-2; (d) PTC(MR).

in turn might lead to overfitting). This is visible on all the datasets, but it is particularly well observable on the COX-2 and PTC(MR) datasets. For example, on the COX-2 dataset, we outperform the reference system by about 30 percentage points during the first two iterations (from 47.3% to 77.4% and from 51.6% to 82.8%, respectively). On PTC(MR), we see a similar pattern for the first three iterations, where our system increases the accuracy with about 20 percentage points (from 41.4% to 62.8%, from 41.4% to 60%, and from 47.1% to 70%). This emphasizes the usefulness of the systematic augmentation by means of matching-graphs, especially when small sets of rather poor training samples are available only.

In Table 1.3, we compare the classification accuracy of our novel system with both reference systems in tabular form. To this end,

Table 1.3. Classification accuracies of two reference systems ($SVM_R(-d_{BP})$ and $SVM_F(-d_{BP})$) and our novel system ($SVM_R+(-d_{BP})$). Symbols ⓧ / ⓨ indicate a statistically significant improvement and ⬤ⓧ / ⬤ⓨ indicate a statistically significant deterioration in x, y of the 20 iterations when compared with the first and second reference system, respectively (using a Z-test at significance level $\alpha = 0.05$).

Dataset	Reference Systems		Ours
	$SVM_R(-d_{BP})$	$SVM_F(-d_{BP})$	$SVM_R+(-d_{BP})$
Mutagenicity	61.8 ± 3.5	69.1	**64.3 ± 2.4** ⑬ / ⬤⑲
NCI1	62.0 ± 3.8	68.6	**64.5 ± 2.4** ⑧ / ⬤⑰
COX-2	65.2 ± 7.8	71.3	**75.6 ± 4.3** ⑪ / ③
PTC(MR)	53.1 ± 6.3	54.3	**64.6 ± 3.4** ⑧ / ⑤

we aggregate the results of the 20 iterations and show the mean classification accuracies (including the standard deviation). We observe that the mean classification accuracy of our approach is better than the first reference system on all data sets. On NCI1, our system outperforms the reference system in all iterations (as seen in Figure 1.2). Eight of the 20 improvements are statistically significant.[f] Eight out of 19, 11 out of 20, and 13 out of 19 improvements are statistically significant on the other datasets, respectively.

On Mutagenicity and NCI1, the novel system does not reach the classification accuracy of the second reference system $SVM_F(-d_{BP})$ that has access to the full training data. Most of the deteriorations are statistically significant. However, we have to keep in mind that it was not our main goal to improve a classifier that has access to the original training set but to show that our approach is able to substantially improve a system that has access to a small dataset only. From this point of view, it is remarkable that for the other two datasets, our novel approach outperforms the second reference system on average. We observe a substantial improvement of about 5 and 10 percentage points on COX-2 and PTC(MR), respectively.

In Table 1.3, it is also visible that the novel system becomes more robust, as the standard deviation is smaller for each dataset

---

[f]The statistical significance is computed via Z-test using a significance level of $\alpha = 0.05$.

for $\text{SVM}_R+(-d_{\text{BP}})$ compared to $\text{SVM}_R(-d_{\text{BP}})$. It is also worth noting that on both Mutagenicity and PTC(MR), the average validation accuracy of our system lies closer to the average test accuracy, compared to the reference system, which might indicate a better generalization of our system.

1.4.3 Graph augmentation for neural networks

The overall aim of the next experimental evaluation (originally described in [26]) is to verify whether augmented sets help make neural network-based classifiers more robust. In order to answer this question, we evaluate the three GNN architectures, described in Section 1.2, with and without data augmentation. That is, the reference models are trained on the full original training sets (denoted by $Simple_F$, GIN_F, and $GraphSAGE_F$), whereas our novel models are trained with matching-graphs added to the training sets (denoted as $Simple_F+$, GIN_F+, and $GraphSAGE_F+$).

In order to counteract uncontrolled randomness of neural network initializations, each experiment is repeated five times and the average accuracy is finally reported (we use the same seeds for both the reference approach and the augmented approach).

1.4.3.1 Validation of metaparameters

For each iteration of the experiment, the corresponding dataset is split into a random training, validation, and test set, with a 60:20:20 split. As we primarily aim at comparing three different network architectures, once with the default training set and once with an augmented training set, we do not separately tune the hyperparameters. Instead, we use the parametrization proposed in [36]. The optimizer used is Adam with a learning rate of 0.001, and we use the cross-entropy as loss function for all three models.

All models are trained for 200 epochs (except for *Mutagenicity* and *NCI1*, where we train for 50 epochs only, due to computational limitations arising from the large number of graphs in these datasets). The models that perform the best on the validation sets are finally applied to the test sets. In this experiment, for each matching-graph that is created, the probability p (Algorithm 1, Line 5) is randomly chosen in the interval $[0.1, 0.9]$.

Table 1.4. Classification accuracies of three models ($Simple_F$, GIN_F, and $GraphSAGE_F$), compared to the same three models with augmented training sets ($Simple_F+$, GIN_F+, and $GraphSAGE_F+$). Symbols ⓧ or ● indicate a statistically significant improvement or deterioration in x of the five iterations when compared with the respective reference system (using a Z-test at significance level $\alpha = 0.05$).

Dataset	$Simple_F$	$Simple_F+$	GIN_F	GIN_F+	Graph-SAGE$_F$	Graph-SAGE$_F+$
Mutagenicity	79.1 ± 0.7	81.7 ± 1.1 ③	80.2 ± 0.9	81.2 ± 0.3 ①	77.6 ± 1.1	77.6 ± 0.7
NCI1	68.1 ± 1.8	71.8 ± 0.7 ⑤	74.0 ± 1.7	76.3 ± 0.8 ③	72.8 ± 0.4	73.1 ± 1.4 ①
COX-2	68.1 ± 4.0	70.9 ± 4.0 ①	73.4 ± 6.0	69.8 ± 1.2 ❷	68.8 ± 3.0	74.3 ± 3.0 ③
PTC(MR)	58.9 ± 7.1	64.0 ± 2.6 ①	62.0 ± 6.4	65.1 ± 1.6 ①	64.7 ± 4.5	62.0 ± 5.5

1.4.3.2 Test results and discussion

In Table 1.4, we compare the mean classification accuracies of the three reference models with the augmented models using matching-graphs (obtained in five iterations).

Overall we observe that the augmentation process generally works well for all three GNN architectures. Using the simple GNN, the augmented approach outperforms the corresponding reference system on all datasets and all iterations. In total, 10 out of the 20 improvements are statistically significant.[g] Using the augmented training sets in conjunction with the GIN_F model, we obtain a higher mean accuracy compared to the reference system on three out of four datasets. In total the augmented model, GIN_F+ statistically significantly outperforms the reference system GIN_F in 4 out of 20 iterations. Using $GraphSAGE_F$ trained on the augmented sets, we outperform the reference system on three out of four datasets according to the mean accuracy. On the *Mutagenicity* and *PTC(MR)* datasets, no statistically significant improvement can be observed, whereas on *NCI1* and *COX-2*, we achieve four statistically significant improvements.

In Figure 1.3, we show — as an example on the *NCI1* dataset — the reference accuracies as well as the accuracies of our approach as bar charts for all five iterations (in light and dark gray, respectively). The iterations are ordered from the worst to the best performing

---

[g]The statistical significance is computed via Z-test using a significance level of $\alpha = 0.05$.

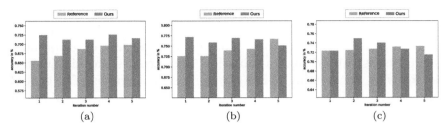

Fig. 1.3. Classification accuracies of the five iterations on the *NCI1* dataset using three different models (light bars) compared to our system that is based on the same models but augments the training set with matching-graphs (dark bars). (a) Simple$_F$; (b) GIN$_F$; (c) GraphSAGE$_F$.

reference accuracy. We can clearly see that our approach outperforms the reference systems for almost all iterations on all models.

Regarding the results in Table 1.4, we also observe that the standard deviation of the accuracies is almost always smaller for the augmented approach (on all datasets). Hence, we can conclude that our system becomes in general more robust and stable when compared with the reference system.

Overall, we observe that the augmentation leads to the least improvements on the *GIN$_F$* model, followed by *GraphSAGE$_F$* and finally *Simple$_F$* with the largest improvements. The differences of the accuracies on the *Simple$_F$* model are the most striking, suggesting that the augmentation approach with matching-graphs helps bridge the gap when no sophisticated network architecture is available or applicable.

1.5 Conclusion and Future Work

Access to sufficient training data is very important and crucial but unfortunately not always given, so in real world applications, one sometimes has to manage with very little labeled data. One possible solution to this problem is to systematically augment the datasets with artificial training data. The process of augmentation is not only interesting in situations where little training data is available but particularly for systems that crucially depend on large training sets (like, for instance, neural network-based classifiers).

In the case of statistical data representations, quite an amount of methods is available for the task of data augmentation. Due to the complex nature of graphs, however, the research of graph augmentation is still behind its statistical counterparts. In this chapter, we focus on augmenting (small) training sets of graphs. The goal of this augmentation process is to make graph-based pattern recognition more robust and ultimately improve the downstream training of a classification algorithms. The novelty of this approach is that we use so-called matching-graphs for the augmentation process.

Matching-graphs, which can be pre-computed by means of (suboptimal) graph edit distance computations, formalize the matching between two graphs. Thereby, they actually define a novel graph that encodes an intermediate representation of two given graphs. In this approach, we systematically produce matching-graphs for each pair of training graphs and are thus able to substantially increase even very small sets of training graphs.

We evaluate the novel procedure in two specific situations and on four graph datasets. First, we research the effects of the novel augmentation process with very small graph sets. By means of an experimental evaluation on artificially reduced graph datasets, we empirically confirm that our novel approach is able to significantly outperform a classifier that has access to the small training set only. Moreover, in some particular cases, we can even report that our novel approach is able to outperform a system that has access to the original set of training graphs. Second, we evaluated the novel approach of augmenting the training data in an experiment with three different graph neural network models. In this scenario, we assume the full data (i.e., we do not artificially reduce the training data). This experiment is particularly interesting because it is known that graph neural networks can be sensitive to the size of the training set. We empirically confirm that our novel approach is able to improve all three GNN architectures in general. The vast majority of the observed improvements is statistically significant. However, our experiments are limited to using datasets with two classes that contain graphs that are relatively small with a low edge density.

For future work, we see different rewarding avenues that can be pursued. First, we feel that it could be beneficial to extend the definition of a matching-graph to include additional nodes and/or edges, to further increase the augmentation capability. Second, it would be interesting to see if the matching-graphs can be used in conjunction with other graph neural networks (e.g., triplet loss networks or others).

References

[1] L. Schoneveld, A. Othmani and H. Abdelkawy, Leveraging recent advances in deep learning for audio-visual emotion recognition, *Pattern Recognition Letters* **146**, pp. 1–7 (2021).

[2] Z. Li, H. Shao, L. Niu and N. Xue, Progressive learning algorithm for efficient person re- identification, in *25th International Conference on Pattern Recognition, ICPR 2020, Virtual Event*, Milan, Italy, January 10–15, 2021. IEEE, pp. 16–23 (2020).

[3] S. Ghosh, S. Ghosh, P. Kumar, E. Scheme and P. P. Roy, A novel spatio-temporal siamese network for 3d signature recognition, *Pattern Recognition Letters* **144**, pp. 13–20 (2021).

[4] F. Battistone and A. Petrosino, TGLSTM: A time based graph deep learning approach to gait recognition, *Pattern Recognition Letters* (Robustness, Security and Regulation Aspects in Current Biometric Systems) **126**, pp. 132–138 (2019).

[5] M. Kenning, J. Deng, M. Edwards and X. Xie, A directed graph convolutional neural network for edge-structured signals in link-fault detection, *Pattern Recognition Letters* **153**, pp. 100–106 (2022).

[6] K. Madi, E. Paquet and H. Kheddouci, New graph distance for deformable 3d objects recognition based on triangle-stars decomposition, *Pattern Recognition* **90**, pp. 297–307 (2019).

[7] D. Conte, P. Foggia, C. Sansone and M. Vento, Thirty years of graph matching in pattern recognition, *International Journal of Pattern Recognition and Artificial Intelligence* **18**(3), pp. 265–298 (2004).

[8] H. Bunke and G. Allermann, Inexact graph matching for structural pattern recognition, *Pattern Recognition Letters* **1**(4), pp. 245–253 (1983).

[9] A. Sanfeliu and K.-S. Fu, A distance measure between attributed relational graphs for pattern recognition, *IEEE Transactions on Systems, Man, and Cybernetics* **13**(3), pp. 353–362 (1983).

[10] K. M. Borgwardt, M. E. Ghisu, F. Llinares-López, L. O'Bray and B. Rieck, Graph kernels: State-of-the-art and future challenges, *Foundations and Trends in Machine Learning* **13**(5–6) (2020).

[11] S. Singh, B. Steiner, J. Hegarty and H. Leather, Using graph neural networks to model the performance of deep neural networks (2021), *CoRR* abs/2108.12489.

[12] M. Fuchs and K. Riesen, Matching of matching-graphs — A novel approach for graph classification, in *25th International Conference on Pattern Recognition, ICPR 2020, Virtual Event*, Milan, Italy, January 10–15, 2021. IEEE, pp. 6570–6576 (2020).

[13] F. Jelinek, Some of my best friends are linguists, *Language Resources and Evaluation* **39**(1), pp. 25–34 (2005).

[14] M. Banko and E. Brill, Mitigating the paucity-of-data problem: Exploring the effect of training corpus size on classifier performance for natural language processing, in *Proceedings of the First International Conference on Human Language Technology Research, HLT 2001*, San Diego, California, USA, March 18–21, 2001. Morgan Kaufmann (2001).

[15] C. Shorten and T. M. Khoshgoftaar, A survey on image data augmentation for deep learning, *Journal of Big Data* **6**, p. 60 (2019).

[16] F. Pereira, P. Norvig and A. Halevy, The unreasonable effectiveness of data, *IEEE Intelligent Systems* **24**(2), 8–12 (2009).

[17] T. Zhao, Y. Liu, L. Neves, O. J. Woodford, M. Jiang and N. Shah, Data augmentation for graph neural networks, in *35th AAAI Conference on Artificial Intelligence, AAAI 2021, 33rd Conference on Innovative Applications of Artificial Intelligence, IAAI 2021, The 11th Symposium on Educational Advances in Artificial Intelligence, EAAI 2021, Virtual Event*, February 2–9, 2021. AAAI Press, pp. 11015–11023 (2021).

[18] J. Zhou, J. Shen, S. Yu, G. Chen and Q. Xuan, M-evolve: Structural-mapping-based data augmentation for graph classification, *IEEE Transactions on Network Science and Engineering* **8**(1), pp. 190–200 (2021).

[19] J. Park, H. Shim and E. Yang, Graph transplant: Node saliency-guided graph mixup with local structure preservation, In *36th AAAI Conference on Artificial Intelligence, AAAI 2022, 34th Conference on Innovative Applications of Artificial Intelligence, IAAI 2022, The 12th Symposium on Educational Advances in Artificial Intelligence, EAAI 2022 Virtual Event*, February 22–March 1, 2022. AAAI Press, pp. 7966–7974 (2022).

[20] H. Zhang, M. Cissé, Y. N. Dauphin and D. Lopez-Paz, mixup: Beyond empirical risk minimization, in *6th International Conference on Learning Representations, ICLR 2018*, Vancouver, BC, Canada, April 30–May 3, 2018, Conference Track Proceedings. OpenReview.net (2018).

[21] T. N. Kipf and M. Welling, Variational graph auto-encoders (2016), *CoRR* abs/1611.07308.

[22] M. Simonovsky and N. Komodakis, Graphvae: Towards generation of small graphs using variational autoencoders, in V. Kurková, Y. Manolopoulos, B. Hammer, L. S. Iliadis and I. Maglogiannis (eds.), *Artificial Neural Networks and Machine Learning — ICANN 2018 — 27th International Conference on Artificial Neural Networks*, Rhodes, Greece, October 4–7, 2018, Proceedings, Part I. Lecture Notes in Computer Science, Vol. 11139. Springer, pp. 412–422 (2018).

[23] H. Wang, J. Wang, J. Wang, M. Zhao, W. Zhang, F. Zhang, X. Xie and M. Guo, Graphgan: Graph representation learning with generative adversarial nets, in S. A. McIlraith and K. Q. Weinberger (eds.), *Proceedings of the 32nd AAAI Conference on Artificial Intelligence, (AAAI-18), the 30th innovative Applications of Artificial Intelligence (IAAI-18), and the 8th AAAI Symposium on Educational Advances in Artificial Intelligence (EAAI-18)*, New Orleans, Louisiana, USA, February 2–7, 2018. AAAI Press, pp. 2508–2515 (2018).

[24] N. De Cao and T. Kipf, Molgan: An implicit generative model for small molecular graphs (2018), *CoRR* abs/1805.11973.

[25] M. Fuchs and K. Riesen, Graph augmentation for small training sets using matching-graphs, *ICPRAI — 3rd International Conference on Pattern Recognition and Artificial Intelligence* (2022).

[26] M. Fuchs and K. Riesen, Graph augmentation for neural networks using matching-graphs, in N. El Gayar, E. Trentin, M. Ravanelli and H. Abbas (eds.), *Artificial Neural Networks in Pattern Recognition — 10th IAPR TC3 Workshop, ANNPR 2022*, Dubai, United Arab Emirates, November 24–26, 2022, Proceedings. Lecture Notes in Computer Science, Vol. 13739. Springer, pp. 3–15 (2022).

[27] X. Chen, H. Huo, J. Huan and J. S. Vitter, Fast computation of graph edit distance (2017), *CoRR* abs/1709.10305.

[28] A. Fischer, C. Y. Suen, V. Frinken, K. Riesen and H. Bunke, Approximation of graph edit distance based on Hausdorff matching, *Pattern Recognition* **48**(2), pp. 331–343 (2015).

[29] K. Riesen and H. Bunke, Approximate graph edit distance computation by means of bipartite graph matching, *Image and Vision Computing* **27**(7), 950–959 (2009).

[30] F. Scarselli, M. Gori, A. C. Tsoi, M. Hagenbuchner and G. Monfardini, The graph neural network model, *IEEE Transactions on Neural Networks* **20**(1), pp. 61–80 (2009).

[31] J. Gilmer, S. S. Schoenholz, P. F. Riley, O. Vinyals and G. E. Dahl, Neural message passing for quantum chemistry, in *International Conference on Machine Learning*. PMLR, pp. 1263–1272 (2017).

[32] Z. Ying, J. You, C. Morris, X. Ren, W. Hamilton and J. Leskovec, Hierarchical graph representation learning with differentiable pooling, *Advances in Neural Information Processing Systems*, Vol. 31 (2018).

[33] C. Morris, M. Ritzert, M. Fey, W. L. Hamilton, J. E. Lenssen, G. Rattan and M. Grohe, Weisfeiler and leman go neural: Higher-order graph neural networks, in *The 33rd AAAI Conference on Artificial Intelligence, AAAI 2019, The 31st Innovative Applications of Artificial Intelligence Conference, IAAI 2019, The 9th AAAI Symposium on Educational Advances in Artificial Intelligence, EAAI 2019*, Honolulu, Hawaii, USA, January 27–February 1, 2019. AAAI Press, pp. 4602–4609 (2019).

[34] K. Xu, W. Hu, J. Leskovec and S. Jegelka, How powerful are graph neural networks? in *7th International Conference on Learning Representations, ICLR 2019*, New Orleans, LA, USA, May 6–9, 2019. OpenReview.net (2019).

[35] W. L. Hamilton, Z. Ying and J. Leskovec, Inductive representation learning on large graphs, in I. Guyon, U. von Luxburg, S. Bengio, H. M. Wallach, R. Fergus, S. V. N. Vishwanathan and R. Garnett (eds.), *Advances in Neural Information Processing Systems 30: Annual Conference on Neural Information Processing Systems 2017*, December 4–9, 2017, Long Beach, CA, USA, pp. 1024–1034 (2017).

[36] F. Errica, M. Podda, D. Bacciu and A. Micheli, A fair comparison of graph neural networks for graph classification, in *Proceedings of the 8th International Conference on Learning Representations (ICLR)* (2020).

[37] C. Morris, N. M. Kriege, F. Bause, K. Kersting, P. Mutzel and M. Neumann, Tudataset: A collection of benchmark datasets for learning with graphs (2020), *CoRR* abs/2007.08663.

[38] N. Wale, I. A. Watson and G. Karypis, Comparison of descriptor spaces for chemical compound retrieval and classification, *Knowledge and Information Systems* **14**(3), pp. 347–375 (2008).

[39] J. J. Sutherland, L. A. O'Brien and D. F. Weaver, Spline-fitting with a genetic algorithm: A method for developing classification structure-activity relationships, *Journal of Chemical Information and Computer Sciences* **43**(6), pp. 1906–1915 (2003).

[40] C. Helma, R. D. King, S. Kramer and A. Srinivasan, The predictive toxicology challenge 2000-2001, *Bioinformatics* **17**(1), pp. 107–108 (2001).

[41] M. Neuhaus and H. Bunke, *Bridging the Gap between Graph Edit Distance and Kernel Machines*. Series in Machine Perception and Artificial Intelligence, Vol. 68. WorldScientific (2007).

[42] K. Riesen and H. Bunke, Classification and clustering of vector space embedded graphs, in *Emerging Topics in Computer Vision and Its Applications*. World Scientific, pp. 49–70 (2012).

© 2025 World Scientific Publishing Company
https://doi.org/10.1142/9789811289125_0002

Chapter 2

Controlling Topology Preserving Graph Pyramids

Walter G. Kropatsch*, Majid Banaeyan*,
and Rocio Gonzalez-Diaz[†]

*TU Wien, Vienna, Austria
[†]University of Seville, Seville, Spain

Abstract

Since the beginning of the use of pyramidal structures for processing images some 40 years ago, several different operations have been applied for a large variety of different applications. The basic advantage of the pyramids is the progressive reduction of the data, level by level by a reduction factor that limits the pyramid's height to the logarithm of the diameter of the base level. Differently from the classical (Gaussian) pyramids, we focus on pyramids where the basic data structure is not an array but a graph structure embedded in the image space. In this chapter, we target (1) topological issues of objects in images like holes in a region, (2) what operations can be used to propagate image information from the input to the high levels as well as in the opposite direction, (3) what specific properties can be generalized by what operations, (4) how to achieve the logarithmic computational complexity, and last but not least, (5) how to coordinate the different processes. In the second part, we focus on a new type of pyramid, the LBP pyramid, that uses a variant of the local binary patterns to recognize critical points and contracts lowest contrast edges during the bottom-up phase. Not only the topology among the relevant parts of the image is preserved, but, as experiments have shown, it also allows the reconstruction of images with only a few colors that are often hard to distinguish from the original.

2.1 Introduction

In this chapter, we describe a hierarchical structure that has its origins in the classical image pyramids like Gaussian or Laplacian pyramids but with the big advantage that the data structure for the individual levels of the pyramid are no more rigid grids or arrays but are based on planar embedded graphs. This enables the pyramid to adapt its structure to the needs of the data: Parts of the data that are considered important for the processing can survive to higher levels while redundancies in the data like homogeneous regions can be reduced during the bottom-up construction phase. Graphs are used here because they are widely known as versatile data structures although the proper representation of topological relations needs the dual graphs. However, there are other less known data structures like combinatorial maps [1] or generalized maps [2] or cell complexes [3,4] that can replace the graphs in the pyramid.

All these data structures do not only describe the topological arrangement of the data but can also describe the complex arrangements of semantic objects of different sizes and shapes that should appear as results of the processing. An important issue in dealing with the huge amounts of data is the possibility to process them in a massive parallel way to reach reasonable processing efficiency. A requirement for parallelism is the independence of the operations such that the result does not depend on the order of the applied operations.

Controlling the big variety of possible choices in the general concept of irregular pyramids is one of the main issues of this chapter. There are several choices in the bottom-up construction of the pyramids but also parameters that have an influence on the abstract concepts surviving to the higher levels need to be chosen or even optimized and adapted to the data. But not only the bottom-up processes are important, irregular pyramids allow also a top-down expansion process that provides an insight into the visual information at the higher levels by, i.e., visualization but also enable to better tune the repeated bottom-up processes with a better overview of what objects with known properties are where in the input. In such cases, attention could be put on particular object details in a repeated bottom-up phase. These up and down phases also provide explanations of what the pyramid has found in its higher levels.

At some places, we also relate the presented concepts to the very popular methods of machine learning (ML) and artificial intelligence (AI). There are differences but also similarities, advantages as well as drawbacks. We see the irregular pyramids not as a competitor of the ML approaches but see possibilities for fruitful combinations.

Section 2.2 starts with five subsections giving motivations and some background of the presented concepts. It follows a recall on irregular pyramids in Section 2.3, the processes for propagating data from the base level to the top as well as in the opposite direction, the expansion from top down to the base. Section 2.4 discusses the many options for controlling the processes, the tasks, and the properties that have been explored in different applications (Section 2.4.5).

In Section 2.5, we adapt the local binary patterns (LBP) to the basic data structure: the graphs, and study their relation to the critical points of curves and surfaces. Monotonic paths, curves, and profiles through an image show that these 1D manifolds have invariant LBPs. On this basis, we construct the LBP pyramid (Section 2.6) and show reconstructions with only a small percentage of the original input. These reconstructions are visually difficult to distinguish from the original data. We draw the conclusion that the structure of the critical points and their adjacencies extracted by the LBP pyramid is extremely important, while the actual gray levels or colors of the image are visually less relevant. This raises the question about the space between the critical points (Section 2.6). We give a simple definition of the concept of a 'slope' with several interesting properties leading to future directions of research addressed in the conclusion.

2.2 Motivations and Background

In this section, we mention five different motivations for the use of irregular pyramids. We start with some requirements described by Leonard Uhr, 1986. We continue with some facts about biological plausibility often used as arguments to justify approaches in recent AI. We then shortly mention a recent project: There we study biological images with extremely high resolutions. The next motivation addresses the problem that not all problems can be solved by the same architecture. Psychology has identified so-called insight problems that cannot be solved by simply optimizing a universal architecture. Finally, we shortly summarize some crucial insights of a seminal paper by Jan Koenderink [6]. They gave us the strong motivation for the research presented in this chapter.

2.2.1 *The problem of biological perception*

Leonard Uhr [5] summarized the problem of human visual perception in 1986 with a few facts and some conclusions: Each human eye has about 10^7 cones and 10^8 rods sensing the light entering the human eye. The measured intensities and frequencies are processed by a large number of synapses where each one takes about 1.5 μs allowing about 1000 serial operations in one second. In order to "see" and to accurately react on the visual stimuli, no more than 600 serial steps are available. This can be achieved by the human brain only by massive parallel processes that converge in logarithmic complexity toward the location where decisions are taken.

Leonard Uhr proposed pyramidal data structures as the only chance to solve the vision problem. But he also clarified that *"pyramids are not (only) multiresolution, parallel bottlenecks, low level, array processors, or trees."* He further stated that *"a pyramid needs augmentation"* and *"... any connected (data-flow) graph could be used."* Furthermore, they need to *"combine bottom-up and top-down"* processes to solve the complex vision problems.

2.2.2 *The human retina is irregular*

Most neural network architectures claim biological plausibility. This is partly true for the general functionality of the signal processing (weighted averages and activation functions), but it certainly does not apply to the underlying architecture: Both the sensors for the visual input as well as the many other sensors providing valuable input for the information processing of the human brain are not regular grids in contrast to most of the artificial neural networks that are currently popular. Figure 2.1 shows a small segment of the

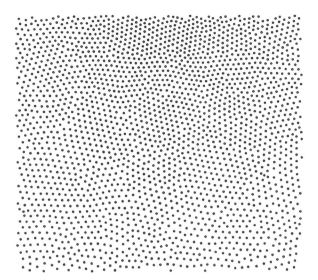

Fig. 2.1. A section of rods and cones in the retina.

retina of a monkey's eye.[a] It is very similar to the human retina and it is clearly not an array! The natural arrangement of sensors in the human eye needs data structures such as graphs to properly represent the irregular embeddings and to learn more about the benefits of these irregular sensor arrangements, in particular the relationship to saccadic eye movements that certainly are not just an accident of nature but may have a considerable importance for the reliable processing of noisy visual data.

2.2.3 *Project: Water's gateway to heaven*

This research project[b] that our group started in 2020 together with two groups in biology raised some very essential problems typical for the trend to use extremely high resolutions and also temporal changes in three dimensions (Figure 2.2).

The project studies 3D imaging and modeling of transient stomatal responses in plant leaves. Input to these studies are

---

[a] Data of the monkey's retina have been gratefully provided by Peter Ahnelt.
[b] https://waters-gateway.boku.ac.at/.

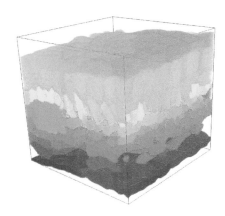

Fig. 2.2.   3D μCT image with color labels.

high-resolution X-ray micro-tomography (μCT) and fluorescence microscopy images. μCT images have the challenging dimensions ranging up to $2000 \times 2000 \times 2000 \approx 2^{33}$ voxels and are taken at 2–4 instances of time. Visible objects are different cells, water ways, and the airspace in between. Leaves are not rigid but to a certain extent deformable. Consequently, rigid matching may not work so well when comparing different images of the same specimen, in particular if the concentration of water is different in the two acquisitions.

The main goal in this project is to **understand** the causality of opening and closing of the stomata. These are cells that can open to allow gases to enter (e.g., CO_2) for **photosynthesis** and water to leave.

The huge amount of data and the complexity of the models describing the processes require a very efficient processing of the data. We are confident that pyramids provide the requested performance.

2.2.4 *Critical/stationary points are relevant*

Jan Koenderink [6] draws some important conclusions in his seminal contribution "The Structure of Images" (1984). He considers intensity images as a function in three-dimensional space $\Phi(x, y, t)$, where (x, y) are the spatial coordinates and t is the scale dimension. He considers the scale as generated by convolution with a Gaussian kernel $\Phi(x, y, 0) * G(t)$. The **Diffusion** $\Delta \Phi = \Phi_t$ is the basis for his scale space theory. He requests that "*Any feature at a coarse resolution is required to possess a 'cause' at finer resolution.*" He considers stationary (critical) points by setting the spatial derivatives to zero: $\Phi_x = \Phi_y = 0$. Among those satisfying these constraints, the Hessian distinguishes between the different critical points:

$$\Phi_{xx}\Phi_{yy} - \Phi_{xy}^2 \geq 0 \text{ for extrema and} \tag{2.1}$$

$$\Phi_{xx}\Phi_{yy} - \Phi_{xy}^2 < 0 \text{ for a saddle point.} \tag{2.2}$$

We shall find a solution in Section 2.5.4 for both decisions **without the noise-sensitive partial derivatives**. A particular observation of Jan Koenderink could be verified after the new identification of critical points: *Extrema and saddle points disappear pairwise when t increases.* It turns out to be useful to eliminate pairs of critical points that are not persistent (i.e., very close peaks with similar height separated by a saddle not much below the peaks).

2.2.5 *An insight problem*

In this fifth motivation, we discuss the limitations of solutions found by optimization processes. In his book, Pizlo [7] demonstrates impressively that there exist problems that cannot be solved simply by optimization (the most frequent strategy for most machine learning approaches). He gives a very simple example:

Create n equilateral triangles ($\triangle$) with m matchsticks:

(1) Create one triangle with three matchsticks (Figure 2.3).
   This has the obvious solution in Figure 2.3.
(2) Make two triangles with two more matchsticks (Figure 2.4).
(3) Can you produce four triangles with one more matchstick?

Fig. 2.3. Three match sticks form one triangle.

Fig. 2.4. Five matchsticks form two triangles.

For the solution, consider the Euler–Poincaré characteristic to balance the number of points (•), the number of matchsticks (m), and the number of triangles (△):

	Euler–Poincaré characteristic					
	#P	-	#E +	#F		= 1
Case	•	-	m +	△		= 1
1.	3	-	3 +	1		= 1
2.	4	-	5 +	2		= 1
3.	?	-	6 +	4		= 1

The last case would suggest that the solution has three points that seems impossible. This is the characteristic of an 'insight problem': An 'insight problem' is typically difficult to solve.

Reference [7] shows an elegant solution with a change in representation: "*If you exclaim 'aha!' at the moment the solution suddenly occurs to you, you had an insight.*" Once the solution strategy is understood, it is easy to explain. However, the above Euler–Poincaré characteristic shows that the optimization would not find a proper solution. A similar reasoning could be applied to several machine learning solutions.

2.3 Recall on Irregular Pyramids

The irregular pyramid consists of a stack of graphs with decreasing size. Each graph of this stack is called a level of the pyramid and the lowest level is the base graph corresponding to the input image where pixels correspond to the graph's vertices and two vertices are joined by an edge if the corresponding pixels are 4-connected. This base graph is also called the neighborhood graph $G(V, E)$ of the image. 4-neighborhood is preferred since edges between diagonal neighbors of 2 × 2 pixels would intersect, with the consequence that the 8-connected graph is not planar. The pixel value is an attribute to the corresponding vertex and it can range from a single gray value to a vector of either spectral channels or additional information like filter responses, lengths, and distances. In order to properly describe the embedding in the image plane, we use the dual graph $\overline{G} = (\overline{V}, \overline{E})$ that is implicitly given by the embedding of the image. The dual vertices $\overline{V}$ identify the face formed by any 2 × 2 block of pixels and the dual edges $\overline{E}$ correspond to the boundary segment between any two adjacent pixels.

2.3.1 Extended region adjacency graph

Image segmentation typically assigns each pixel a label identifying the set of pixels having the same or a similar property. The adjacencies of these regions are typically described by the region adjacency graph (RAG) where each vertex represents a connected set of pixels with the same label and two vertices are connected in the RAG if two regions with two different labels share a common boundary.

Most approaches consider the RAG as a simple graph without multiple edges and without self-loops. But the simple graph cannot describe all the topological configurations that these regions can be related to in practice: The left and right riverbanks of a river may be connected by more than one bridge. The simple RAG just states that the two riverbanks are connected but not by how many bridges. This can be resolved by simple RAGs by sub-dividing each riverbank into as many segments as there are bridges. This is not only increasing the size of the graph, but it is also difficult to handle since the characteristic features of the segments may be similar if not identical such that they cannot be easily classified.

A second example where the simple RAG has problems describes the relationship between a lake and its islands. The outer boundary of the lake is a closed curve and each island is also bounded by a closed curve: In a typical inclusion relationship, the islands are completely surrounded by the lake. Let us describe the mainland with a vertex of the RAG, the lake with a vertex, and each island also with a vertex. Clearly, the mainland is connected to the lake and the lake is connected to each of its islands. But what expresses the fact that the lake surrounds all islands? One solution is to introduce a separate data structure, an inclusion tree. It works in 2D, but what about a tunnel in 3D?

We found the extension of the simple RAG, a good solution to solve both problems: The multiple bridges can be represented by multiple edges without the need to arbitrarily sub-divide the homogeneous riverbanks and self-loops that surround the islands can represent the inclusion relation. To distinguish the more frequently used RAG from the non-simple RAG, we denote the extended version by E-RAG.

2.3.2 Overview of the bottom-up construction

Figure 2.5 gives an overview of irregular pyramids. The base level is the 4-neighborhood graph of the image and each level above the base represents an E-RAG $G = (V, E)$.

The next higher level is reached by contracting selected edges while preserving certain relevant points. They form the **contraction kernel**. The smaller graph contains less vertices and less edges, but some edges have become multiple and some even self-loops. Therefore, the next step is to **simplify** the graph from unnecessary multi-edges and self-loops. Before repeating the contraction, the attributes of the newly generated, smaller graph need to be derived by **reduction functions** taking as input the receptive field of each surviving vertex and edge and computing the attributes of the elements of the higher-level graph. Then, this process can be repeated until a **termination criterium** is satisfied and the apex of the pyramid is reached. The overall process is controlled by the following steps:

- the selected contraction kernel,
- the simplification process,
- the reduction function, and
- the termination criterium.

Fig. 2.5. Bottom-up and top-down processes in an irregular pyramid.

If the reduction process reduces the graph from level to level by a constant **reduction factor** ≥ 2, then the height of the pyramid is bound by the logarithm of the diameter of the base graph. This contributes to the efficiency of the pyramid when the level by level processing can be massively parallel (compare with Uhr [5] and Section 2.2.1).

2.3.3 *Overview of the top-down reconstruction*

The apex graph of the pyramid is a very abstract representation of the visual entities of the image and their spatial and topological relations. For the purpose of explaining what has been derived from the given input image, the high levels can be successively down-projected to the lower levels and to the base in order to show the entities that have been derived above. For this purpose, we keep some information about the bottom-up process that enables then to reverse the construction and to propagate downwards the insights gained at the higher levels hopefully explaining what and why certain entities have been found.

The basis for the reconstruction is the canonical representation of Torres and Kropatsch [8]. It stores the contraction kernels and simplification parameters in chronological order together with links that enable to undo edge contraction by edge decontraction, edge removal by edge reinsertion, and the attributes at input for the reduction function.

The following subsections introduce more details about these processes with the purpose of showing some interesting properties.

2.3.4 *Contracting an edge*

Definition 2.1 (Edge contraction). The operation of contracting an edge $e = (v_1, v_2) \in E, v_1 \neq v_2 \in V$, of a graph $G = (V, E)$ consists in first identifying the two end points $v_1 \mapsto v_s, v_2 \mapsto v_s$ of the edge e into a new 'surviving' vertex $v_s \in \{v_1, v_2\}$ and replacing v_1, v_2 in all edges by v_s. Finally, the edge e is removed.

The graph after the contraction of edge e has one less edge and one less vertex: $G' = G/e = (V \setminus \{v_1, v_2\} \cup \{v_s\}, E \setminus \{e\})$. Note that the condition $v_1 \neq v_2$ excludes self-loops (v, v) from being contracted.

Contraction preserves the connectivity of G in G'. As the dual operation of contracting an edge in G/e, the corresponding dual edge is removed from $\overline{G}$: $\overline{G'} = \overline{G} \setminus \overline{e}$. Consequently, the dual graph of G' needs only the dual operation applied to $\overline{G}$ and the duality is preserved.

As a result of contraction, G' may contain parallel edges and even self-loops. Most of them can be removed in the successive simplification step in Section 2.3.6. The remaining parallel edges and self-loops identify special topological properties like the inclusion of holes.

Independent edges can be contracted simultaneously in parallel. All edges that are simultaneously contracted form a **contraction kernel**.

2.3.5 *Contraction kernel*

In order to be able to execute many contractions in parallel (with many processors), they must be independent of each other. In other words, the order in which the set of edges is contracted should not affect the result. Several methods have been used to create contraction kernels (CK) with independent edges: maximal independent vertex set (MIS), Meer [9], maximal independent edge set (MIES), and maximal independent, directed edge set (MIDES), Kropatsch et al. [10] with different properties and advantages.

Definition 2.2 (Contraction kernel). Let $G(V, E)$ be the input graph to be contracted. A contraction kernel $K \subset E$ is a subset of edges that forms a spanning forest of G. Each tree of the forest contains one surviving vertex; in some extreme cases, the tree can even be a single (surviving) vertex.

There may be different criteria (examples are given in Section 2.4) for selecting concrete edges to contract and for selecting vertices to survive. The surviving vertices are the vertices of the next pyramid level; the edges are the result of the contraction processes. If each connected component of the contraction kernel covers at least two vertices of V_n, the number of vertices V_{n+1} will be less than $|V_n|/2$. We call this the reduction factor of 2. Isolated vertices can be compensated by larger trees in different parts of the graph. Both selection methods, MIES and MIDES, have this property. If the trees of the contraction kernel are independent of each other, all can be contracted in parallel, while the edges of each tree may need sequential processing. The most efficient contraction kernels are many small trees with more than one vertex.

Consequently, if the height of the pyramid has h levels, then the base graph has $|V_0| \geq 2^h$ vertices. If all the trees of the forest are independent and a sufficient number of processors are available, the next pyramid level can be computed in $\mathcal{O}(\max\{\deg(v)|v \in V_n\})$ parallel steps.

2.3.6 *Simplifying multiple edges and self-loops*

The contraction of one edge of a triangle leads to the creation of a double edge or even multiple edges. The contraction of one of the multiple edges creates self-loops (Figure 2.6). Note that the dual faces f_1, f_2 are preserved. Before the first two contractions, the degree of the faces $\deg(f_1) = \deg(f_2) = 3$. Inside the triple edges, the degrees shrink to 2 and the self-loops surround faces with degree 1. This example also shows that a simplification after the first contraction would simplify the further processing.

Multiple edges and self-loops are not topology relevant if they don't surround any further (sub-)structure. This can be decided by looking at the dual graph $\overline{G}(\overline{V}, \overline{E})$ where the degree of a face $\in \overline{V}$ provides such a decision:

Definition 2.3 (Topology-relevant). A face of the dual graph $\overline{G}$ is topology relevant for G if its degree is higher than 2: $\deg(\overline{v}) > 2$ for $\overline{v} \in \overline{V}$.

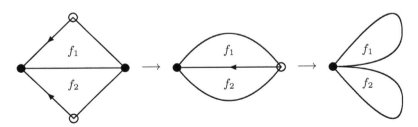

Fig. 2.6. Creation of multiple edges and self-loops.

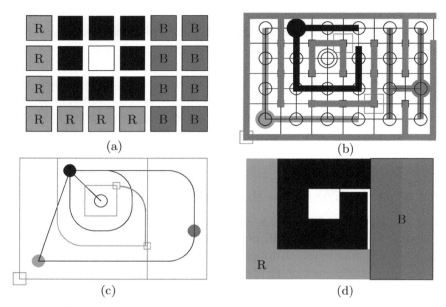

Fig. 2.7. A pseudo-edge connects the white island. (a) 4×6 colored pixels; (b) equivalent CK and RK; (c) resulting E-RAG and $\overline{E - RAG}$; (d) 4 colored regions with pseudo-edge.

Multiple edges and self-loops surrounding topology-irrelevant faces are not relevant for topology; self-loops can be removed without disconnecting either a hole or any sub-structure in the dual graph. Multiple edges can be removed as well, as long as the last remaining edge is preserved to keep the connectivity. The remaining edges are called pseudo-edges since they have the same face on both sides (see the example in Figure 2.7). Let us denote all the edges that can be removed as the removal kernel (RK).

Definition 2.4 (Removal kernel). Let $G'(V', E') = G(V, E)/K$ be the graph after contracting all edges of the contraction kernel K and let $\overline{G'}$ be its dual:

$$R'(G') = \{(v,v) \in E' | f \in \overline{(v,v)} \subset \overline{V'}, \deg(f) = 1\}, \tag{2.3}$$
$$\cup \ \{e_1 = e_2 = (v,w) \in E' | f = \overline{e_1} \cap \overline{e_2} \subset \overline{V}, \deg(f) = 2\}. \tag{2.4}$$

The set of edges in R' can be removed without modifying the topology relevance of the graph. Note that removing the edges of the removal kernel may create further redundant edges. For the complete simplification, a few more iterations of simplification may be needed, since the removal of an edge may decrease the degree of adjacent faces and may create further edges that can be removed. The complexity of this process has been shown to be the inverse of the Ackermann function [11]. A faster version has been proposed by Banaeyan and Kropatsch [12] by anticipating the contractions and removing the redundant edges in parallel before actually executing the contractions.

2.3.7 Example with a hole and a pseudo-edge

Edges that are not relevant for topology are often called **redundant**. There is one exception: if the removal of an edge would disconnect the graph or its dual graph. Consider the example in Figure 2.7(a). It shows the 24 pixels with the colors red, black, blue, and white. The white pixel is completely surrounded by the black connected component. Figure 2.7(b) shows the contraction kernels for the three colors red, black, and blue together with the selected surviving vertices. The white pixel survives and is indicated in Figure 2.7(b) by two concentric circles. The removal kernels are shown in green with surviving dual vertices (these are the intersections of the boundaries) marked by green squares. The background is the larger square in the left bottom corner.

Figure 2.7(c) shows the pair of dual graphs after contracting and simplifying the CK and RK. The fact that the black region completely surrounds the white pixel is expressed by the self-loop attached to the black vertex in $G(V, E)$. The edge dual to this self-loop is the pseudo-edge which connects the boundary of the white pixel with the intersection of the three connected components of the red, black, and blue regions. We call it "pseudo"-edge since both sides have the same color black while all other dual edges have different colors on both sides. The geometric placement of the pseudo-edge can be any connection of the boundary of the white with any intersection of the black region with other colors. It is illustrated by the white line in Figure 2.7(d).

The main role of the pseudo-edge is to keep the graph $\overline{G(V,E)}$ connected. The pseudo-edge is a bridge in $\overline{G(V,E)}$, the dual of which expresses the fact that black surrounds white.

If there are multiple holes in a region, each hole creates one pseudo-edge. Since their placements just need to cross the surrounding region, both can connect to the outer boundary of the surrounding region or, equivalently, only one connects to the outer boundary and the other connects the two holes. Together with the pseudo-edges the surrounding region remains homeomorphic to a topological ball. And reversely, each pseudo-edge indicates the presence of a hole in 2D. Extensions to higher dimensions exist but are not treated here.

2.3.8 *The bottom-up construction of the irregular pyramid*

The bottom-up construction of an irregular pyramid is an iterative parallel process that can be repeated until all the properties to be transferred bottom-up are application-relevant and any further shrinking would destroy relevant properties or relations. This process generates an abstraction of the base-level graph.

Given graph $G_0(V_0, E_0)$ and its dual graph $\overline{G_0}(\overline{V_0}, \overline{E_0})$, iteration count $n = 0$.

While further abstraction is possible do

(1) select contraction kernels $K_n \subset E_n$ as in Definition 2.2;
(2) perform contraction $G' = G/K_n$, $n = n + 1$;
(3) select removal kernel $R'(G')$ as in Definition 2.4;
(4) and simplify $G_n = G' \setminus R'$;
(5) apply reduction functions $RF(\cdot) : G(K_{n-1}) \to$ new reduced content

$\text{attr}(v_n) = RF(\mathcal{N}_V(v_{n-1})), v_n \in V_n$ and
$\text{attr}(e_n) = RF(\mathcal{N}_E(e_{n-1})), e_n \in E_n$.

Each iteration creates a new level $G_i(V_i, E_i), i = 0, \ldots n$, of the pyramid.

2.3.9 Preserving topology

Already in Section 2.2.5 we used the Euler–Poincaré characteristic referring to the relationship between the number of points P, of edges E, and of faces F in a 2D plane graph. Let us now consider the changes Δ created by the primitive operations, edge contraction, and edge removal:

Change of Euler–Poincaré characteristic

Operation	$\Delta\#P$	$-$	$\Delta\#E$	$+$	$\Delta\#F$	$= 0$
Contraction	1	$-$	1	$+$	0	$= 0$
Removal	0	$-$	1	$+$	1	$= 0$

That means that the characteristic does not change after the application of our primitive operations. More generally, any number of **contractions and removals do NOT change the characteristic!**

We have seen that regions surround their holes by a self-loop, the dual of which is a pseudo-edge. By keeping these pseudo-edges, the characteristic of the region is not changed since together with the pseudo-edges the region remains homeomorphic to a topological ball.

2.3.10 Equivalent contraction kernels

Similar to the equivalent weighting functions in Burt's regular pyramid, Ref. [13] introduces equivalent contraction kernels. Contraction kernels cover the receptive field of the surviving vertex. For every edge e_n in a higher pyramid level n, there exists one edge $e_i, 0 \leq i < n$, in the levels below that survives to e_n in the sense that if e_n is contracted at level $n+1$, then edge e_0 can be added to the contraction kernel $K_0(v_{n+1})$ at the base level such that the receptive field is covered by $K_0(v_{n+1})$ for vertex v_{n+1}. With the same argument, the equivalent contraction kernel of the top vertex is a spanning tree of the receptive field in the base level of the pyramid.

2.3.11 The top-down expansion process

Top-down expansion has been used effectively in classical Laplacian pyramids by Burt and Adelson [14]. In regular pyramids, the structure of the pyramid depends only on the size of the base image and hence the size of the different levels above the base does NOT vary for images with the same input size. For irregular pyramids, the structure of the graphs of the different levels depends strongly on the content of the data. The selection, the contracted edges, as well as the other control parameters may depend on the content of the image. Hence, irregular pyramids on different images may have a completely different graph structure. However, we built them bottom-up, level by level, and with only two different operations: contraction and removal of edges.

In Ref. [8], we have shown (1) that there are inverse operations to the two basic operations and (2) that we need to remember only a few parameters of the bottom-up process to reconstruct the higher-resolution graph. We call the inverse operations **decontraction** of a contracted edge and **reinsertion** of a removed edge (Figure 2.8). In this canonical encoding of the irregular pyramid, we store the parameters of the contracted and removed edges in the order they have been applied. These recycled garbage parameters allow us in the top-down reverse process to recover the graphs at the lower levels.

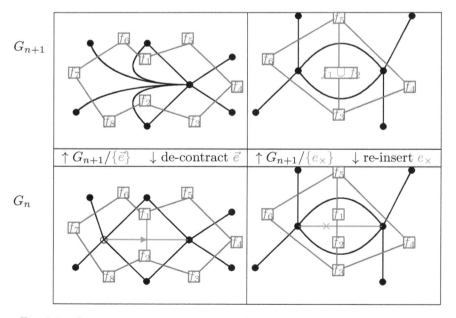

Fig. 2.8. Inverse operations: contract and de-contract, remove, and reinsert.

The canonical encoding enables first of all to reconstruct the levels below the top level. However, more importantly, we can down-propagate the abstract attributes collected in the higher levels. Different options are discussed in Section 2.4.4.

2.4 Control by the Content

The previous section covered the main components for constructing an irregular pyramid and for expanding the abstract information from the top level down to the pixels of the original image. We can identify four categories of control over the general process, influencing either the constructed structure or the architecture of the hierarchy or preserving certain real-world properties of objects to be represented in the base-level image: The selection of contraction kernels identifies the surviving vertices and some of the incident edges that are not relevant for the main properties of the objects. The simplification strategies 'clean' the graph after each contraction phase. Also here there are possible choices, e.g., what parallel edges should survive. Reduction functions use the attributes of the survivor's children to compute a more abstract description of the content of the receptive field. Once a certain number of levels of the pyramid has been generated, the extracted high-level description can be expanded to the lower levels in order to (1) display the abstract content of any higher level in the form of an image and (2) probably revise some decisions taken at the bottom-up process in order to make the content of the complete pyramid consistent with the abstract findings at the higher levels.

2.4.1 *Select contraction kernels*

The simplest choice of contraction kernels is a random choice for constructing the stochastic pyramid of Meer [9]. In the adaptive pyramid of Jolion and Montanvert [15], the random choice is replaced by choosing the irregular sampling from the content of the data. Such adaptation could be convolution filters of which a local maximum identifies the surviving vertex. Note the high similarity to common '*max-pooling*' layer in deep learning architectures. But also rules can be used to select edges to be contracted. In connected component

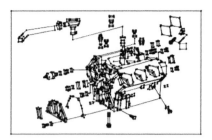

Fig. 2.9. Technical drawing of a motor engine.

labeling (CCL), a simple rule is to contract only edges connecting vertices with the same label. A more complicated rule for selecting contraction kernels has been used for closing gaps in scanned line drawings [16, 17] (see the example in Figure 2.9). Even parametric models could be used for determining the important vertices to preserve and the edges to contract. This could be as simple as correlating the data with the model or finding the best *'goodness of match'*. Finally, the matching of graphs that is in general NP hard could be done using the fact that using the same selection rules for two images is likely to generate much simpler and similar graphs at higher levels of the two pyramids in this case [18].

2.4.2 *Simplification strategies*

There are two different criteria for selecting the removal kernels: either the content-based choice in choosing the surviving edge of multiple edges according to the attributes of the edge (e.g., shortest accumulated arc length) or the attributes of the two adjacent faces of the dual graph (e.g., distance to the outer most parallel edge) or the computational choice of how many iterations of simplification should be done. Complete simplification after each contraction needs $\mathcal{O}(a^{-1}(n,n))$ steps in the worst case where $a(n,n)$ is the Ackermann function. Alternatively, only one simplification pass is executed after contraction, leaving the remaining multiple edges and self-loops for simplification at higher levels. This may of course indirectly slow down the construction since neighborhoods with non-simplified redundant edges are larger. The last alternative is to do all simplifications after all contractions.

Under certain conditions (having a total order of all vertices), simplification can be anticipated before contraction.

A noteworthy alternative has recently been proposed by Banaeyan [19]. This approach can achieve simplification prior to contraction under certain conditions, specifically when a total order of all the vertices exists. In the case of a binary image, independent edges (i.e., edges not sharing an endpoint, [20]) are encoded to allow for the removal of redundant edges originally at an upper level, at the current level with parallel constant complexity [12]. This method accelerates the construction of the pyramid and transforms it into an efficient tool for computing the distance transform of a binary image with parallel logarithmic complexity [21], provided that there is sufficient number of processing elements for parallel computations.

2.4.3 Reduction functions

The role of reduction functions is to propagate the image content to a lower resolution while at the same time increasing the degree of abstraction. While a pixel in the base may have the color, i.e., red, it may be aggregated at a higher level into a red ball.

The simplest reduction function is used in CCL: All the vertices in the contraction kernel have the same color hence the surviving vertex will inherit the same color. The second most frequently used choice is a (weighted) average or, more generally, a convolution filter (as frequently used in DCNN^c). A more sophisticated reduction uses the transitive closure of a set of relations (i.e., describing the layout of curves in line drawings, such as Figure 2.9 [16]). Both in the processing of line drawings as in the closing of gaps [22], i.e., between the dashes of a dashed line, the introduction of an **isolated blob** ⊙ allows establishing neighbor relations between the dashes rather than connecting all the dashes to the common (white) background. Figure 2.10 shows an example: the input image, the resulting graph, and the receptive fields of the different isolated blobs. The survivor received an additional symbol ⊙ for dashed/dotted lines if the black dash • appears completely surrounded by the white background ○. The rule for contraction is then extended by the isolated blob ⊙ : in addition to the fact that the same categories • , ○ can be merged as

---

^{c}Deep Convolutional Neural Network.

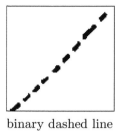 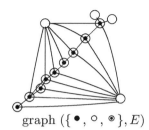 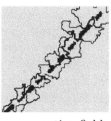

binary dashed line   graph $(\{\bullet, \circ, \odot\}, E)$ receptive fields

Fig. 2.10. Recognizing a dashed line.

in CCL, we allow $\odot$ to merge with $\circ$ but not with $\odot$. The growth of the isolated blobs happens concurrently to the growth of the (empty) background $\circ$ such that close-by $\odot$ neighbors are detected before all the background merges into a large region where the individual blobs are all surrounded by individual self-loops.

But also parametric models may determine the parameters best describing the receptive field of the surviving vertex. Of course, models can become more complex and parameters that best match the data [23] can be used to describe the vertex by the name of the model and its parameters. All these models offer opportunities for optimization and learning.

There is no need to use the same reduction function when reducing one level to the next. Of course, it is the simplest choice if no other source of information is available. But if you consider the dynamic processing of visual data or have a target segmentation available, there may be previous labels and features available such that the reduction function can be adapted for the general model from the previous image frame. And not only concerning the parameters of the reduction function but also the principle type of function, e.g., switching from a filter to inheritance or the transitive closure of the boundary segments.

2.4.4 *Controlling the top-down expansion*

As with the classical Laplacian pyramid [14], a first motivation is to show that the original image can be reconstructed from the higher levels. But even more, a simple inheritance expansion where children inherit their attributes from its parents, without trying to reconstruct the original attributes, provides some insight into what has been aggregated in the higher levels. In the Laplacian expansion, the high frequencies of the lower levels have not been added.

In the irregular pyramid, the expansion has become feasible by the inverse operations [8], decontraction for contraction and reinsertion for removed edges. Originally, these inverse operations were applied in reverse order to be able to re-establish properly the links to the already expanded graphs. However, this strict order, which would prevent parallel application, can be relaxed since also the bottom-up operations were independent and create layers of contracted edges (by one contraction kernel) alternating with removed edges through simplification. Similar to the concept of wavelets, this process can be memory neutral in the sense that the active level where the current top level graph is stored complements the passive part where links of the contracted and removed edges are kept. Together they occupy the same memory as the base level.

The recovery of structure of the lower levels of the pyramid offers a wide variety of possibilities to propagate high-level information (referred to as the parents, the surviving vertex together with its neighbors) to the lower levels (referred to as the children). Options that have been used are as follows:

- interpolating the attributes of the children from the attributes of the parents,
- or using convolution filters applied on the parent's level,
- or inheriting the parent's attributes (as for CCL),
- or refining the high-level model and potentially updating the bottom-up model by properties like straightness of a dotted line that cannot be done locally during the bottom-up process,
- or reinsertion of curve segments to re-establish connectivity,
- or generative models like fractals.

We give examples for some of the operations in the following section.

2.4.5 *Preserving relevant properties*

In Table 2.1, we give examples of the different choices of the control decisions used by specific applications together with citations to papers with the details and results. In nearly all cases, empty faces,

Table 2.1. Overview of control for specific applications.

Application	Important elements survive	Negligible elements are merged
CCL	1 repr/CC(lab)	(L, L)
segmentation	1 repr/ region	similar, end points
2x on curve	X, ends	empty space, connections
line images	ends, junctions	empty space, connections
matching	discrim.template, object boundary	simil.inside object
motion	foreground, static background, articulations	occluded backgr. moving foregr.
gap closing	1 repr/lab incl. background	(L, L)
E-RAG Hierarchy	max.ext.Contrast, MST	min.int.Contrast

with deg < 3, are considered redundant in the simplification and merged with one adjacent face (corresponding to the removal of the separating edge).

Connected component labeling (**CCL**) [17, 24–26] has as input a labeled image. It could be a hand-labeled ground truth or the result of a segmentation, and the task is to find the connected components of the different labels together with their adjacencies. One vertex of every connected component should survive to the top (1 repr/CC(lab) in Table 2.1) while edges connecting vertices with the same label can be contracted (denoted by (L, L) in Table 2.1).

There are numerous studies of **segmentation**, i.e., [27–30]. In this case, every connected region will be represented by one surviving vertex in the top level and edges connecting similar vertices are contracted to the edges of the RAG. For thin regions, it may be useful to keep the end points to some higher levels.

The psychological test of "**2X on a curve**" consists in finding out whether two X placed on two complicated but non-overlapping curves are on the same curve or on different curves. It was argued that humans need a time proportional to the length of the curves. In our paper [31], we showed that the pyramid can solve it in logarithmic time by (1) preserving the "X" vertices and (2) contracting the empty space without curve segments and contracting connected curve segments.

In processing **line images** (e.g., technical drawings, Figure 2.9), the preservation of line ends and of junctions is important [32–34]. Similar to the previous application, contraction applies to the empty space and to connected curve segments. For line images, there was an additional constraint that the face should not touch any curve because it would establish a wrong connectivity. Here the adaptivity of the irregular pyramids is a great advantage.

In the application of finding **matchings** between two images [18, 35, 36], as in stereo or in image mosaics, the most discriminative template should survive together with the object boundaries while edges connecting similar vertices inside an object can be contracted.

The problem of detecting **motion** in image sequences involves more than a single or a pair of images [37–39]. In this application, the task is to identify a moving object in front of a static background and to identify the moving parts of an articulated movement (walking or hand gestures). It is important to keep one vertex of each connected foreground object and the static parts of the background. In addition, the articulation points need to be preserved in order to derive, e.g., a proper walking pattern. Contraction can be applied to edges inside the background or inside a moving foreground object. Expansion can be used to build a more complete background model by inserting parts that have been temporarily occluded by a moving foreground as well as tracking the moving foreground objects over time.

The **gap closing** application [40, 41] has been discussed together with the drawing of line images in Section 2.4.3. We have shown that the introduction of a new label ⊙ for isolated blobs can be determined locally by the given graph structure (the self-loop surrounding a blob) and can be efficiently used as new entity to control the growth of the different categories of labels.

The last example in Table 2.1 is entitled "**E-RAG hierarchy**" [42–44]. The preservation of topology enables the classical region adjacency graph to allow also self-loops and multiple edges. These are necessary to properly represent the inclusion of holes in a large region and the fact that two regions may be connected by more than one connected boundary. The criteria used in this application were that vertices with the highest external contrast (according to [45]) survive and the edges of a minimal spanning tree of the internal contrast are contracted.

The last application, the LBP-pyramid, is discussed in Section 2.6.

2.4.6 *Properties of topological pyramids*

Let us call topological the pyramids that preserve the topological properties of the data/images. In particular, it concerns holes of regions and the related inclusion relationships. Topological properties are to a large extent invariant to geometric deformations like different view points, perspective projection, articulated movements, etc. But topological properties are also sensitive to noise and care must be taken when removing noise.

Concerning the data structure for storing the topological pyramid, the matrix structure of a regular pyramid is definitely not able to properly represent all relevant topological features explicitly. For example, a small hole may quickly be too small to be represented at lower resolutions. But more importantly, thin structures like roads or rivers in a remotely sensed image are likely to disappear when their width drops below the sampling distance. That is why we have focussed on embedded graphs as a primary data structure, although there are less known representations like combinatorial maps, generalized maps, or cellular complexes that suit the purpose for preserving topology as well.

We already addressed an important aspect of graphs: Simple graphs without multiple edges and self-loops cannot capture holes and multiple connected boundaries. We showed that graph pyramids can be constructed with only two operations: edge contraction and edge removal.

We have seen also the particular importance of preserving key vertices to higher levels; they allow keeping the overview of the main components of an image and often relax particular details that may not be necessary once the object has been identified.

In two dimensions, plane graphs represent the graph embedded in the plane as shown in Figure 2.11. The base of the primal graph corresponds to the 4-neighborhood of the image while the dual graph has an important role in deciding the removal kernels without long search processes.

The bottom-up construction is controlled by application-specific properties. It preserves the connectivity and the relevant inclusions

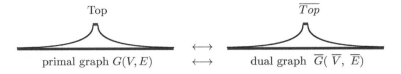

Fig. 2.11. Dual graph pyramid.

during the bottom-up process. Pseudo-edges are bridges in the dual graph that connect a hole with the remaining graph. Their deletion would disconnect the graph and remove the information about what end point is included in the receptive field of the other. Its dual edge is a self-loop indicating the inclusion. Each hole can be associated with one pseudo-edge. This remains true also in higher dimensions, i.e., a pseudo-face characterizes a tunnel through a volume, a typical example is a torus.

The concept of equivalent contraction kernel (ECK) relates the higher levels to the lower levels directly without the need to propagate across several levels. The $ECK(v)$ of any vertex of the pyramid covers the complete receptive field of vertex v. This becomes of particular interest for color images where each color channel creates a separate pyramid structure. In this case, it is very difficult to compare the higher levels of the three pyramids directly. However, through the ECK, each vertex can be down-projected to the common image structure where the comparison could be done.

Another important aspect of pyramids is that many operations can be executed in parallel on different processors. We have shown ways to identify independent operations, making the computational complexity even for large images as those mentioned in Section 2.2.3 feasible [19].

2.5 Local Binary Patterns (LBPs)

Local binary patterns have been introduced by Ojala and Pietikainen [46] in 1996 as an efficient descriptor for textures in images. The eight neighbors of the center of a 3×3 window compare their gray value with the center and set a 1 if the neighbor is higher in value and a 0 otherwise. The resulting eight bits are concatenated in a pre-defined (clockwise) order and form a value in [0,255]. This works very well for the eight neighbors of an 8-connected grid of an image.

It fails if the number of neighbors varies like in a graph with vertices of different degrees. However, a graph has also edges in addition to the vertices. We therefore store the result of the comparison not with the (center) vertex but with the edge connecting the center vertex with the neighbor by simply orienting the edge such that it always points to the lower valued vertex. In this case, the characteristic bit switches of LBPs translates into an orientation switch of the edges surrounding a vertex. This way not only relaxes the degree of the vertices but also saves more than 50% of the memory. In addition, all the characteristics of LBPs like the differentiation of critical points (minima, maxima, and saddle points) translate 1-1 to the new representation.

LBPs identify the class of uniform codes: These are codes that contain maximally two bit-switches when turning around the center. In our new representation, these are local configurations that are either extrema (0 bit switches) or their neighborhood splits into a higher connected part and a lower connected part separated by a level curve across the center. The two bit-switch configurations roughly form a slope (precise definition is given in the following). Non-uniform LBPs correspond to saddle points.

2.5.1 *Critical points of a height profile*

Let us first consider an LBP along a one-dimensional (1D) curve (Figure 2.12) or a profile across a two-dimensional (2D) surface. Basic mathematics tells us that critical points are characterized by horizontal tangents. In 1D, critical points are local extrema: local maxima $\oplus$ and local minima $\ominus$. From Figure 2.12, we see that the curves

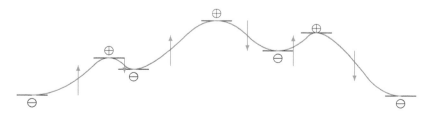

Fig. 2.12. A smooth curve with seven critical points.

between the critical points are **monotonically** increasing from the minima toward the maxima while they **monotonically** decrease from the maxima toward the minima.

Definition 2.5 (Monotonic). A function $f(x) : D \mapsto \mathbb{R}$ is called **monotonically increasing** in the domain $D \subset \mathbb{R}^n$ if $f(y) \leq f(x)$ for all $x \leq y$ in D, and it is called **monotonically decreasing** in the domain $D \subset \mathbb{R}^n$ if $f(x) \leq f(y)$ for all $x \leq y$ in D. It is called **strictly monotonic** for strict inequalities.

2.5.2 *LBPs along a monotonic curve*

LBPs compare a central point $f(x)$ with its neighbors $\mathcal{N}(x) = \{n | \delta(n, x) \leq \Delta\}$. In 1D, we use $\delta(x, y) = |x - y|$ and $\Delta = 1$, and in 2D, the Euclidean distance

$$\delta\left(\begin{pmatrix} x_1 \\ y_1 \end{pmatrix}, \begin{pmatrix} x_2 \\ y_2 \end{pmatrix}\right) = \sqrt{(x_1 - x_2)^2 + (y_1 - y_2)^2}. \qquad (2.5)$$

LBP stores a binary value of 0 if the neighbor is smaller or equal and a value of 1 if the neighbor is greater than the central point:

$$LBP(x) = b_0, b_1 \text{ with} \qquad (2.6)$$

$$b_i(x) = \begin{cases} 0 \ldots & \text{iff } f(n_i) \leq f(x) \\ 1 \ldots & \text{iff } f(n_i) > f(x) \end{cases} \text{ and } i \in \{0, 1\}. \qquad (2.7)$$

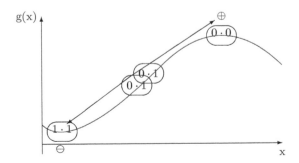

Fig. 2.13. A 1D curve between a local minimum and a local maximum.

In 1D, every point has two neighbors: one (n_0) with lower x and one (n_1) with higher x (Figure 2.13). Consequently, there are four different LBP codes:

code	meaning
00	local maximum ($\oplus$)
01	monotonically increasing curves
10	monotonically decreasing curves
11	local minimum ($\ominus$)

2.5.3 *Monotonic curves/paths* π

Figure 2.14 shows a curve with two sharp peaks (local maxima) and a flat minimum. In the following, it shows the corresponding graph $G(V, E)$ where V represents the critical points/segments ($\oplus, \ominus$) of the curve and the edges between are oriented such that the end point is lower than the begin of the corresponding curve segment. Along the monotonic curves, the orientation of all the edges remains the same, hence they can be collapsed into a single edge if a varying steepness does not matter. The critical points in a 1D continuous curve can be determined by a 1D LBP without computing derivatives and even at non-smooth locations (like the two sharp peaks in Figure 2.14). At sharp corners, there are multiple orientations of tangents and the derivative cannot be computed.

The curve segments between the critical points correspond to the edges and they are all oriented toward the minimum: $\oplus \longrightarrow \ominus$ is monotonically decreasing and $\ominus \longleftarrow \oplus$ is monotonically increasing. Note that the curve need not be smooth with the only requirement that the sampling satisfies the Nyquist–Shannon theorem (at least for the critical points [47]).

A discrete monotonic path $\pi(p_1, p_n) = (p_1, \ldots, p_n)$, $p_i \in \mathbb{R}^n$, is a polygon in $\mathbb{R}^n$ without self-intersection. Formally, we can state

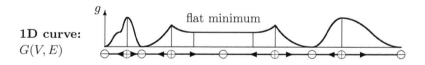

Fig. 2.14. A 1D curve with its oriented graph $G(V, E)$.

that the orientation of a monotonic sequence of edges $(p_i, p_{i+1}) \in E$ can be derived from the sign of σ in (2.8):

$$(f(p_{i+1}) - f(p_i))\sigma \leq 0 \quad \forall i \in [1, n-1], \sigma \in \{-1, +1\}. \tag{2.8}$$

The sequence is increasing with $\sigma = +1$ and decreasing with $\sigma = -1$.

The original LBP bits associated with the vertices in V are transferred to the orientation of the respective edges. Turning around any vertex $v \in V$ in a plane graph,[d] we can derive the corresponding LBP: bit 0 if the edge is $e = (v, w) \in E$ and 1 for $e = (w, v) \in E$. The change in orientation corresponds to a bit switch in the LBP code and enables vertices with different degrees.

Flat regions introduce an asymmetric LBP code in the original definition of Ojala *et al.* [46] giving raise to a local ternary pattern (LTP) in the work of Tan and Triggs [48]. The drawback is that 2 bits are necessary instead of 1 bit doubling the size of the code and there is no obvious translation into orientation. Since flat edges do not contribute to the detection of critical points, we propose to contract flat edges in the graph. This preserves the connectivity and converts monotonicity into strict monotonicity. There is one exception: self-loops are crucial for describing holes in a region. But self-loops are easy to detect by the fact that both end points are the same vertex. As we explore later, a small extension to edges with the smallest contrast allows removing most flat edges. As a side effect, it also shortens the paths π and preserves the critical points.

2.5.4 *Critical points in 2D*

Critical points in 2D can be recognized by **LBP** [46]. As in 1D, bit switches translate into a change in orientation of the edges between adjacent neighbors. We keep the downwards orientation as in 1D. Hence, there are no changes in the orientation of an extremum and two changes for saddle points (see examples in Figure 2.15). Even a third category of local configuration with two bit switches can be described by 'uniform' LBP codes: slopes.

---

[d]A plane graph is an embedded planar graph such that the order of edges around every vertex in the embedding is given.

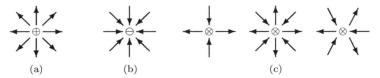

Fig. 2.15. Examples of critical points in 2D. (a) local max. $\oplus$; (b) local min. $\ominus$; (c) local saddle $\otimes$ with degrees 4, 8, 6.

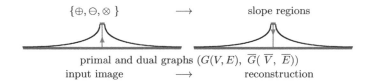

Fig. 2.16. Bottom-up and top-down processes in the LBP pyramid.

2.6 The LBP Pyramid

This section follows the general principles laid out in Section 2.3. References [49, 50] focus (see overview in Figure 2.16) on the particular choices of the LBP pyramid. We have seen in the previous section that the binary coding of local binary patterns (LBP) can be transferred to an oriented graph and that the LBP along a monotonic path shows the same pattern, e.g., the orientation of edges along a monotonic path is the same. Critical points can be determined from the number of orientation changes around a vertex, in most cases by a local process. In case of flat regions, the detection of critical points is postponed until all the flat regions are represented by a single vertex. Then the detection is local.

There is one special case for a 'hidden' saddle point. This is a saddle point that falls between the sampling points of the data and is characterized by a condition related to the non-well-composed 2×2 configurations of Latecki [51]:

$$
\begin{array}{c}
\begin{array}{cc} | & | \\ --A----B-- \\ | & | \\ | \times | \\ | & | \\ --D----C-- \\ | & | \end{array}
\quad \longrightarrow \quad
\begin{array}{cc} | & | \\ --A----B-- \\ |\searrow\swarrow| \\ | \otimes | \\ |\swarrow\searrow| \\ --D----C-- \\ | & | \end{array}
\end{array}
\qquad (2.9)
$$

The block of pixels A, B, C, D is not well composed if either

$$L = \max(A, C) < H = \min(B, D) \text{ or} \qquad (2.10)$$
$$L = \max(B, D) < H = \min(A, C). \qquad (2.11)$$

In all such cases, an extra saddle vertex $\otimes$ is inserted in the center of the 2×2 block of the neighborhood graph, connected to all four vertices A, B, C, and D of the 2×2 pixels by edges and with a gray value $f(\otimes)$ in the interval (L, H).

2.6.1 *Bottom-up construction and top-down expansion*

As contraction kernels, we select edges with the locally lowest contrast and choose the critical points as survivors. If an edge with lowest contrast is not incident to a critical point, any incident vertex can be chosen. There are two main arguments for this choice:

(1) Since low-contrast edges are visually nearly indistinguishable, we use the following selection criterium for edges to be contracted: Contracting edges of contrast zero shrinks successively flat areas until reaching a single vertex for each connected flat area.[e] After zero-contrast edges have been contracted, the remaining edges with contrast $> 0, e = (v, w) \in E$ are downwards-oriented, i.e., $f(v) > f(w)$, and the contrast of e is the difference between the end points: $contrast(e) = f(v) - f(w)$. In order to satisfy the independence condition, one can first determine a maximal independent vertex set starting with the critical points and then choose trees of incident edges with locally lowest contrast.

(2) Each connected component of the contraction kernel contains one critical point, if possible. It remains critical even after contraction, and, consequently, the critical point survives the contraction process. Preserving the critical points $\oplus, \ominus, \otimes$ of the base graph

---

[e]Care must be taken if these flat areas connect non-connected parts of the boundary of the graph since the remaining vertex becomes an articulation point the removal of which would disconnect the graph. This can be avoided by first contracting and simplifying the boundary and then the inner flat areas while preserving the boundary. This enables treating sub-graphs separately and stitching them together after contraction.

does not shrink the range of values (in contrast to smoothing or interpolation). Together with a strong contrast they contribute to the high visual quality of the reconstructed image. In case of dense clusters of critical points or in case of critical points generated by noise, the rule of preservation may be relaxed for generalization by allowing lowest contrast pairs $(\oplus, \ominus), (\oplus, \otimes), (\otimes, \ominus)$ to be contracted (as in [6]).

The attributes of the base level are the gray values of the pixels. An alternative would be to use the contrast as an attribute of an edge where the orientation encodes the sign of the contrast. With this encoding, only a few gray values need to be kept with some vertices since the other gray values can be recomputed by propagation along the edges of the graph.

In the simplification of multiple edges, the longest equivalent path in the base can be chosen. This length can be easily integrated in the bottom-up process by first initializing a length attribute of each edge by the value 1, and after each edge contraction, the length of the edge becomes the sum of the lengths of the two involved edges.

As reduction function, surviving vertices inherit the value of the level below. Since critical points are primarily chosen for survival, the range of gray values is preserved.

The top-down expansion can be done using the canonical representation [8] by edge decontraction and (removed) edge reinsertion. During the expansion process, the children of the lower level either retrieve their value from the status of the bottom-up process or they inherit the value of their parent from the level above. In the experiments, we chose the option of inheritance to judge the quality of the reconstruction with only a few values of the top level.

2.6.2 *Main properties of the LBP pyramid*

The bottom-up construction preserves relevant critical points (see Table 2.2) that are determined by LBP [50, 52, 53]. Hidden saddle points are inserted in the original neighborhood graph.[f] The selection of edges with the lowest contrast also preserves the original contrast

---

[f]This is easy in a graph but would require an increase of the resolution of an array.

Table 2.2. Control of LBP pyramid.

Application	Important elements survive	Negligible elements are merged
LBP pyramid	critical points, texture, high freq.	lowest contrast

of the image as well as the quality of fine details like the grass in Figure 2.18. Monotonic paths remain monotonic if no critical point is removed. LBP codes are known for their texture representation and this property is clearly visible in the reconstructions.

The reconstruction quality of the LBP pyramid has been tested in Refs. [50,52,53] with a variety of images from the Berkeley database. The reconstructions use much less colors and preserve very well the structure and topology of the image. Selected pictures in Figure 2.17 are from the Berkeley image database [54]. More examples can be found in PRIP TR-133, the Master Thesis of Martin Cerman.[g]

2.6.3 Image = Structure + Few Colors

In Ref. [55], we investigated the reasons why the LBP pyramid reconstructs images with surprisingly high visual quality. We could confirm the main observation of Koenderink [6], although he used Gaussian-type smoothing for the construction of the lower scales that cannot preserve so well thin structures as the LBP pyramid does. This can be visually verified in Figure 2.18 where three reconstructions with the LBP pyramid are compared with a classical Gaussian pyramid with a reduction window of 5×5 pixels and a reduction factor of 4, corresponding to a stride of 2 in x and y directions. GE stands for one Gaussian reduction and one expansion, and GGEE for two Gaussian reductions and two expansions. Figure 2.18 summarizes the number of vertices or pixels at the apex of the pyramid and the corresponding reduction from the base to the top. Most of the critical points survive to the low resolutions at high levels and contracting the lowest contrast first preserves the high contrast in the image. And, in contrast to ALL smoothing reductions involving convolutions, it preserves high

---

[g]https://www.prip.tuwien.ac.at/publications/technical_reports.php.

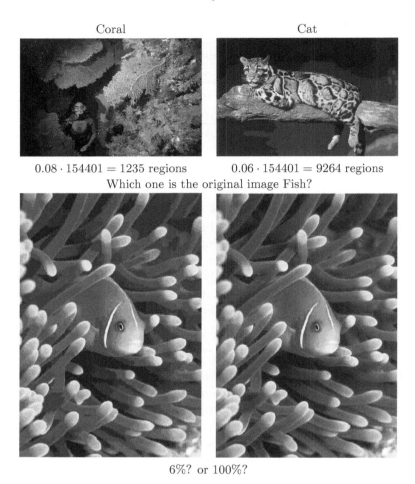

Fig. 2.17. Three reconstructions with the LBP pyramid.

frequencies in the image (small, thin details). Reconstructions with only a few highest levels give good results. Reconstructions with only 30% down to 3% of the regions of the original number of pixels are shown in Figure 2.18.

In Ref. [55], we qualitatively and quantitatively compared the results on 100 images of the Berkeley database with 4 other methods. For the assessment of the quality, we used the Structural Similarity Index Measure (SSIM) [56], the Feature Similarity Index Measure (FSIM) [57], and the Peak Signal-to-Noise Ratio (PSNR).

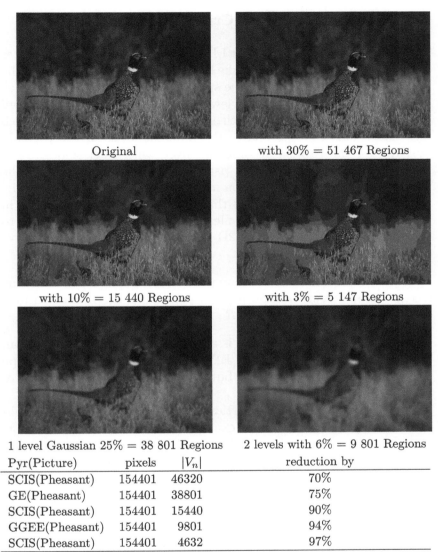

Pyr(Picture)	pixels	$|V_n|$	reduction by
SCIS(Pheasant)	154401	46320	70%
GE(Pheasant)	154401	38801	75%
SCIS(Pheasant)	154401	15440	90%
GGEE(Pheasant)	154401	9801	94%
SCIS(Pheasant)	154401	4632	97%

$|V_n|$ is the number of vertices/pixels at top
'reduction by' is the reduction of the vertices from the base to the top level.

Fig. 2.18. Pheasant, Berkeley# 43074, three reconstructions, two 5 × 5/4 Gaussian.

2.7 The Space between Critical Points

After constructing an irregular LBP pyramid, most of the vertices correspond to critical points of the base level. Hence Ref. [58] asks the following question: What are the spaces between the critical points? In this section, we give an overview about the concept of **slopes** and their interesting properties with some outlook for future research directions. This concept relates the different levels of the pyramid by covering the regions at different levels with such slopes. Since the critical points determine a partition into slopes, it enables the interpretation of the information at different levels of the pyramid.

We have seen in Figure 2.13 that the curves between extrema in 1D are monotonic curves or profiles. Now, we extend this concept to two dimensions. We define the 2D counterpart of a monotonic curve as a slope:

Definition 2.6 (Slope). A connected region R of a continuous surface is a **slope region** iff all pairs of points $\in R$ are connected by a continuous monotonic curve $\in R$.

The smallest slope in an image is a single pixel. Locally, a slope can be characterized by a uniform LBP.[h] Definition 2.6 defines the slope in continuous Euclidean space, but it is also valid in discrete spaces like images or graphs.

We recall some of its properties and refer to our previous publications for proofs and further examples. The domain of the image function f can be partitioned by slopes. Critical points determine the structure of a slope $\mathcal{S}$: Every slope can contain one local maximum ($\oplus$) and one local minimum ($\ominus$). Saddles ($\otimes$) appear exclusively on the boundaries between slopes [59]. All **level curves**[i] in a slope are connected. Level curves of f may be open when intersecting the boundary of the domain of f or closed. Level curves can intersect exclusively at saddle points, never inside R (more in [60]).

We distinguish between two types of slopes:

- slopes bounded by level curves and
- slopes bounded by monotonic curves.

---

[h]A uniform LBP has maximally two bit switches or, equivalently, maximally two changes of orientation when turning around the center.
[i]Also called contour lines or isolines.

2.7.1 Slopes bounded by level curves

The complete boundary is a level curve at the level $f(\boxtimes)$ of the saddle point[j] with following constraints if a local maximum $\oplus$ and/or a local minimum $\ominus$ exists inside the slope (Figure 2.19(a)):

$$f(\oplus) > f(\boxtimes) > f(\ominus). \tag{2.12}$$

The region including the $\oplus$ will be higher than the boundary, and the region with $\ominus$ will be below the boundary if both minimum and maximum are inside the slope. This implies that there exists a curve inside the slope at the level $f(\boxtimes)$ that separates the higher from the lower parts of the slope. This level curve connects the two saddle points $\otimes$ having the same level. These two saddle points of identical level are necessary to prevent the slope from having an articulation point.

Both the boundary and the separation curve meet at the two saddle points. All other level curves inside the slope are closed. Orienting level curves such that the right side is higher than the left side, the level curves opposite $\boxtimes$ have opposite directions (see Figure 2.20).

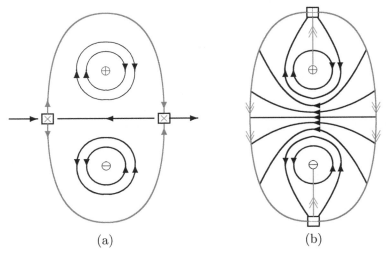

Fig. 2.19. Two basic types of slopes. (a) A level-bounded slope; (b) A slope bounded by two monotonic curves.

---

[j]$\boxtimes$ denotes a saddle point along the boundary, $\boxplus$ and $\boxminus$ a local maximum and a local minimum along the boundary, while $\oplus, \ominus$ denote local extrema in 2D.

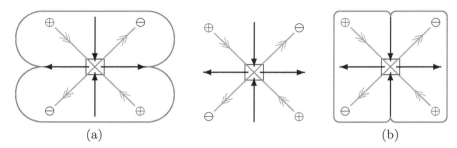

Fig. 2.20. Orientations around a saddle point.

Each extremum inside S is surrounded by closed level curves. Regions around ⊠ alternate higher and lower parts of f (Figure 2.20; ⇐ indicates that the edge increases from left to right). As a consequence, there are two ways to group the higher and the lower parts into two slopes with maximal extension: (a) using horizontal neighbors in Figure 2.20 or (b) using vertical neighbors. Such grouping can be related with certain semantic properties like size, shape, or texture. It can be established at the top level of the pyramid and then successively refined top-down, level by level, to the base level.

2.7.2 Slope with a monotonic boundary

This is most likely the more frequent type of slope since the bounding saddle points have different levels and are connected by monotonic curves. Figure 2.19(b) shows a prototype of such a slope. We draw boundary curves in green, and monotonic curves with ⇐ pointing downwards the levels.

Level λ curves inside a slope are connected and can be open or closed with the following constraints:

$$\overset{\ominus}{[} \quad \overset{\Longleftarrow}{\text{closed}} \quad \overset{\boxminus}{](} \quad \overset{\Longleftarrow}{\text{open}} \quad \overset{\boxplus}{)[} \quad \overset{\Longleftarrow}{\text{closed}} \quad \overset{\oplus}{]}$$

Level curves around the two extrema are closed for levels λ satisfying:

$$f(\boxplus) \leq f(\lambda) \leq f(\oplus), \tag{2.13}$$

$$f(\ominus) \leq f(\lambda) \leq f(\boxminus). \tag{2.14}$$

All level curves $f^{-1}(\lambda)$ with $f(\square) < f(\lambda) < f(\boxplus)$ are open and connect the two monotonic branches of the boundary (except $\square, \boxplus$).

Slopes can partition any continuous surface [59]. However, partitioning into slope regions is not unique. Monotonic boundaries have some degree of flexibility to grow or shrink within the limits of the level curves through the saddle points. This opens also the possibility to introduce a limited overlap between two slopes where the boundary is not clearly determined.

2.7.3 *Outlook on slopes*

Besides the receptive fields of high level vertices and faces, slopes provide a further tool to explain the derived structure and attributes of the higher levels at the higher resolutions in the levels below. One target could be to derive a covering of the base level with a minimal number of slopes. Together with features derived for each slope it could be used to recognize similar image regions or objects in the image.

Since saddle points appear exclusively along the boundary of slopes and the level curves inside have characteristic patterns, we could consider the level curves that pass through saddle points. The connected components between these level curves form a hierarchical structure that has great similarity with the topological tree of shapes [61]. These similarities would be interesting to study not only due to the continuous and the discrete concept but also where the two concepts match and whether there are differences.

Finally, any hill-climbing inside a slope region reaches the peak, and any steepest descent inside a slope region reaches the minimum in all cases. There is definitely a potential for optimization processes to avoid being trapped in intermediate local extrema.

2.8 Conclusion

In this chapter, we first give an overview of the main components of irregular pyramids. They differ from the classical (regular) pyramids in that they are based on irregular data structures like graphs. This enables them to adapt their internal structure to the input data in the base level or to the target structures important for the application.

Irregular pyramids preserve the intrinsic (cell) structure[k] at higher levels. In this chapter, we have used plane graphs as basic topological data structures because graphs can be assumed to be widely known. But several other data structures can be used with varying advantages, in particular for dimensions higher than two: combinatorial maps [1], generalized maps [2], or CW complexes [3,4].

Irregular pyramids have a strong biological motivation: They satisfy the biologically motivated architectural and functional requirements of Uhr [5]. The basis of an irregular pyramid need not be an array but any irregular graph as the Delaunay-triangulated [62] human retina (or any subgraph of it). Also, the neural connections in the brain differ strongly from the artificial counter parts that are currently very popular; they are neither fully connected bipartite graphs nor regular local connections. However, there is some similarity in the functionality of the connections with the significant difference that natural neurons are much slower than massively parallel architectures or modern GPUs. This efficiency of the parallel architecture has become very important in the project 'Water's gateway to heaven' where the 2^{33} data can be processed in parallel up and down the only 33 levels of the irregular pyramid.

The irregular pyramids accept as input an image, data from a retina, any plane graph, combinatorial map, or generalized map. The main goal is

to reduce the huge amount of data while preserving certain properties.

The construction and expansion in irregular pyramids is controlled by

- selected contraction and removal kernels,
- massive parallel graph contraction,
- different reduction functions (decoupled from the flexible architecture!),

---

[k]The term "cell" is just a coincidence with the biological cells in Section 2.2.3, here an abstract cell is meant.

- inverse operations of contraction and removal: decontraction and reinsertion, and
- termination criteria.

Let us repeat here that the choice of reduction functions can vary not only between different levels of the pyramid but also within the same level in the case that there is a strong hypothesis that a certain object with specific properties is located at a particular location. In this case, a concept similar to the object-oriented programming paradigma can be applied.

For the LBP pyramid, we first translate the binary LBP code into the orientation of edges. This enables the recognition of critical points $\oplus, \ominus, \otimes$ and slopes without the need of derivatives. This follows the principle of changing the representation to solve an insight problem (Section 2.2.5). The observation that continuous curves between extrema are monotonic inspired the choice of the lowest contrast for contracting edges. Experimental results showed that in images the arrangement of critical points plays a dominant role and images can be reconstructed with only a few colors if the structure is preserved. Finally, monotonicity also led in 2D to the novel concept of a slope, opening possibilities to explain the achieved results through integrating bottom-up and top-down processes and the relations between the receptive fields at different levels of the pyramid.

In contrast to many machine learning approaches, like deep convolutional neural networks, which have some architectural similarities with the pyramid but have a strong association between the layer and the applied functionality (i.e., convolution layers and pooling layers), irregular pyramids separate their architecture and functionality. In addition, the construction of the hierarchy enables the architecture to adapt the representation to the structure of the data.

The algorithms operating in the irregular pyramid are designed to work (also) on massively parallel architectures with a parallel complexity of $\mathcal{O}(\log(\text{diameter}))$ following parent–child links.

Acknowledgement

Part of this chapter was supported by the Vienna Science and Technology Fund (WWTF), project LS19-013.

Contributions by...

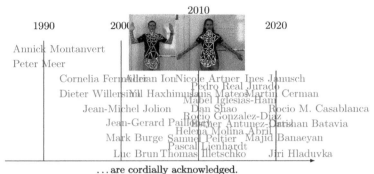

...are cordially acknowledged.
New, future collaborations are very welcome!

References

[1] L. Brun and W. G. Kropatsch, Introduction to combinatorial pyramids, in G. Bertrand, A. Imiya and R. Klette (eds.), *Digital and Image Geometry*. Lecture Notes in Computer Science, Vol. 2243. Springer, Berlin, pp. 108–128 (2001).

[2] G. Damiand and P. Lienhardt, Removal and contraction for n-dimensional generalized combinatorial maps, in H. Wildenauer and W. G. Kropatsch (eds.), *Computer Vision — CVWW'02, Computer Vision Winter Workshop*. PRIP, TU Wien, Wien, Austria, pp. 208–221 (2002).

[3] R. Forman, Morse theory for cell complexes, *Advances in Mathematics* **134**, pp. 90–145 (1998).

[4] R. Forman, Combinatorial differential topology and geometry, *New Perspectives in Geometric Combinatorics* **38**, pp. 177–206 (1999).

[5] L. Uhr, Parallel, hierarchical software/hardware pyramid architectures, in V. Cantoni and S. Levialdi (eds.), *Pyramidal Systems for Image Processing and Computer Vision*. NATO ASI Series, Vol. F25. Springer-Verlag, Berlin, pp. 1–20 (1986).

[6] J. J. Koenderink, The structure of images, *Biological Cybernetics* **50**, pp. 363–370 (1984).

[7] Z. Pizlo, *Solving Solving: Cognitive Mechanisms and Formal Models*. Cambridge University Press, Cambridge (2022).

[8] F. Torres and W. G. Kropatsch, Canonical encoding of the combinatorial pyramid, in Z. Kúkelová and J. Heller (eds.), *Proceedings of the 19th Computer Vision Winter Workshop 2014*. Křtiny, CZ, pp. 118–125 (2014).

[9] P. Meer, Stochastic image pyramids, *Computer Vision, Graphics, and Image Processing* **45**(3), pp. 269–294 (1989).

[10] W. G. Kropatsch, Y. Haxhimusa, Z. Pizlo and G. Langs, Vision pyramids that do not grow too high, *Pattern Recognition Letters* **26**(3), pp. 319–337 (2005).

[11] D. Willersinn, *Irreguläre Kurvenpyramiden: ein Schema für perzeptuelle Organisation*, Ph.D. thesis, Vienna University of Technology (1995).

[12] M. Banaeyan and W. G. Kropatsch, Parallel $\mathcal{O}(\log(n))$ computation of the adjacency of connected components, in *3rd International Conference on Pattern Recognition and Artificial Intelligence*. LNCS, Vol. 13364. Springer (2022).

[13] W. G. Kropatsch, Equivalent contraction kernels to build dual irregular pyramids, *Advances in Computer Vision*. Advances in Computer Science, pp. 99–107 (1997).

[14] P. J. Burt and E. H. Adelson, The Laplacian pyramid as a compact image code, *IEEE Transactions on Communications* **31**(4), pp. 532–540 (1983).

[15] J.-M. Jolion and A. Montanvert, The adaptive pyramid, a framework for 2D image analysis, *Computer Vision, Graphics, and Image Processing: Image Understanding* **55**(3), pp. 339–348 (1992).

[16] M. Burge and W. G. Kropatsch, A minimal line property preserving representation of line images, *Computing, Devoted Issue on Image Processing* **62**, pp. 355–368 (1999).

[17] M. Banaeyan and W. G. Kropatsch, Pyramidal connnected component labeling by irregular pyramid, in *5th International Conference on Pattern Recognition and Image Analysis*. IEEE (2021).

[18] J.-G. Pailloncy, W. G. Kropatsch and J.-M. Jolion, Object matching on irregular pyramid, in A. K. Jain, S. Venkatesh and B. C. Lovell (eds.), *14th International Conference on Pattern Recognition*, Vol. II. IEEE Computer Society, pp. 1721–1723 (1998).

[19] M. Banaeyan, D. Batavia and W. G. Kropatsch, Removing redundancies in binary images, in *2nd International Conference on Intelligent Systems & Pattern Recognition*. LNCS. Springer (2022), doi: 10.1109/IPRIA53572.2021.9483533.

[20] M. Banaeyan and W. G. Kropatsch, Fast labaled spanning tree in binary irregular graph pyramids, *Journal of Engineering Research and Sciences* **1**(10), pp. 69–78 (2022).

[21] M. Banaeyan, W. G. Kropatsch and J. Hladůvka, Fast distance transforms in graphs and in gmaps, in A. Krzyzak, C. Suen and A. Torsello (eds.), *S+SSPR 2022*. Lecture Notes in Computer Science, Vol. 13813. Springer Nature, pp. 193–202 (2022).

[22] W. G. Kropatsch, Building irregular pyramids by dual graph contraction, *IEE-Proceedings Vision, Image and Signal Processing* **142**(6), pp. 366–374 (1995).

[23] T.-H. Hong and A. Rosenfeld, Compact region extraction using weighted pixel linking in a pyramids, *IEEE Transactions on Pattern Analysis and Machine Intelligence* **6**(2), pp. 222–229 (1984).

[24] W. G. Kropatsch and H. Macho, Finding the structure of connected components using dual irregular pyramids, in *Cinquième Colloque DGCI*. LLAIC1, Université d'Auvergne, pp. 147–158 (1995).

[25] H. Macho and W. G. Kropatsch, Finding connected components with dual irregular pyramids, in F. Solina and W. G. Kropatsch (eds.), *Visual Modules*, OCG-Schriftenreihe, Österr. Arbeitsgemeinschaft für Mustererkennung, band 81. R. Oldenburg, pp. 313–321 (1995).

[26] W. G. Kropatsch, Properties of pyramidal representations, *Theoretical Foundations of Computer Vision*. Computing Supplement, Vol. 11, pp. 99–111 (1996).

[27] W. G. Kropatsch and S. BenYacoub, Universal segmentation with p*IRR*amids, in A. Pinz (ed.), *Pattern Recognition 1996, Proceedings of 20th ÖAGM Workshop*, OCG-Schriftenreihe, Österr. Arbeitsgemeinschaft für Mustererkennung, band 90. R. Oldenburg, pp. 171–182 (1996).

[28] W. G. Kropatsch and S. BenYacoub, A general pyramid segmentation algorithm, in R. Melter, A. Y. Wu and L. Latecki (eds.), *Vision Geometry V, International Symposium on Optical Sciences, Engineering, and Instrumentation*, Vol. 2826. SPIE, pp. 216–224 (1996).

[29] W. G. Kropatsch and Y. Haxhimusa, Grouping and segmentation in a hierarchy of graphs, in C. A. Bouman and E. L. Miller (eds.), *Computational Imaging II*, Vol. 5299. SPIE, Bellingham, pp. 193–204 (2004).

[30] A. Hanbury, J. Marchadier and W. G. Kropatsch, The redundancy pyramid and its application to image segmentation, in W. Burger and J. Scharinger (eds.), *Digital Imaging in Media and Education, 28th ÖAGM Workshop*, OCG-Schriftenreihe, Österr. Arbeitsgemeinschaft für Mustererkennung, band 179. R. Oldenburg, pp. 157–164 (2004).

[31] W. G. Kropatsch, Property preserving hierarchical graph transformations, in C. Arcelli, L. P. Cordella and G. Sanniti di Baja (eds.), *Advances in Visual Form Analysis*. World Scientific Publishing Company, pp. 340–349 (1998).

[32] W. G. Kropatsch, M. Burge and H. L. Idl, Dual graph contraction for run graphs, in A. Leonardis and F. E. Solina (eds.), *Computer Vision — CVWW'98, Proceedings of the Computer Vision Winter Workshop*. IEEE Slovenia Section, Ljubljana, pp. 75–86 (1998).

[33] M. Burge and W. G. Kropatsch, Contracting line images using run graphs, in M. Gengler, M. Prinz and E. Schuster (eds.), *22nd ÖAGM Workshop on Pattern Recognition and Medical Computer Vision 1998*, OCG-Schriftenreihe, Österr. Arbeitsgemeinschaft für Mustererkennung, band 106. R. Oldenburg, pp. 235–244 (1998).

[34] W. G. Kropatsch and M. Burge, Minimizing the topological structure of line images, in A. Amin, D. Dori, P. Pudil and H. Freeman (eds.), *Advances in Pattern Recognition, Joint IAPR International Workshops SSPR'98 and SPR'98*, Sydney, Australia. Lecture Notes in Computer Science, Vol. 1451. Springer, Berlin, pp. 149–158 (1998).

[35] R. Glantz, M. Pelillo and W. G. Kropatsch, Matching hierarchies of segmentations, in H. Wildenauer and W. G. Kropatsch (eds.), *Computer Vision — CVWW'02, Computer Vision Winter Workshop*. PRIP, TU Wien, Wien, Austria, pp. 149–158 (2002).

[36] R. Glantz, M. Pelilo and W. G. Kropatsch, Matching segmentation hierarchies, *International Journal for Pattern Recognition and Artificial Intelligence* **18**(3), pp. 397–424 (2004).

[37] R. Glantz, R. Englert and W. G. Kropatsch, Contracting distance maps of pores to pore networks, in N. E. Brändle (ed.), *Computer Vision — CVWW'99, Proceedings of the Computer Vision Winter Workshop*. PRIP TU Wien, Wien, Austria, pp. 112–121 (1999).

[38] J. Marchadier, W. G. Kropatsch and A. Hanbury, The redundancy pyramid and its application to segmentation on an image sequence, in C. E. Rasmussen, H. H. Bülthoff, M. A. Giese and B. Schölkopf (eds.), *DAGM Symposium 2004*, Tübingen, Germany. Lecture Notes in Computer Science, Vol. 3175. Springer, Berlin, pp. 432–439 (2004).

[39] N. M. Artner, A. Ion and W. G. Kropatsch, Rigid part decomposition in a graph pyramid, in J. O. E. Eduardo Bayro-Corrochano (ed.), *The 14th International Congress on Pattern Recognition, CIARP 2009*. Lecture Notes in Computer Science, Vol. 5856. Springer-Verlag, Berlin, pp. 758–765 (2009).

[40] W. G. Kropatsch, Abstraction pyramids on discrete representations, in A. Braquelaire, J.-O. Lachaud and A. Vialard (eds.), *Discrete Geometry for Computer Imagery, 10th DGCI*, Bordeaux, France. Lecture Notes in Computer Science, Vol. 2301. Springer, Berlin, pp. 1–21 (2002).

[41] W. G. Kropatsch and Y. Haxhimusa, Hierarchical grouping of non-connected structures, in W. Burger and J. Scharinger (eds.), *Digital Imaging in Media and Education, 28th ÖAGM Workshop*, OCG-Schriftenreihe, Österr. Arbeitsgemeinschaft für Mustererkennung, band 179. R. Oldenburg, pp. 165–172 (2004).

[42] Y. Haxhimusa and W. G. Kropatsch, Hierarchy of partitions with dual graph contraction, in E. Michaelis and G. Krell (eds.), *DAGM 2003, 25th DAGM Symposium*, Magdeburg, Germany. Lecture Notes in Computer Science, Vol. 2781. Springer, Berlin, pp. 338–345 (2003).

[43] Y. Haxhimusa and W. G. Kropatsch, Segmentation graph hierarchies, in A. Fred, T. Caelli, R. P. Duin, A. Campilho and D. de Ridder (eds.), *Structural, Syntactic, and Statistical Pattern Recognition, Joint IAPR International Workshops on SSPR 2004 and SPR 2004*, Lisbon, Portugal. Lecture Notes in Computer Science, Vol. 3138. Springer, Berlin, pp. 343–351 (2004).

[44] Y. Haxhimusa, A. Ion and W. G. Kropatsch, Comparing hierarchies of segmentations: Humans, nomalized cut, and minimum spanning tree, in F. Lenzen, O. Scherzer and M. Vincze (eds.), *Proceedings of 30th OEAGM Workshop*, Obergurgl, Austria. OCG-Schriftenreihe books@ocg.at, band 209. Österreichische Computer Gesellschaft, pp. 95–103 (2006).

[45] P. F. Felzenzwalb and D. Huttenlocher, Image segmentation using local variation, in *Computer Vision and Pattern Recognition (CVPR)*. IEEE, pp. 98–104 (1998).

[46] T. Ojala, M. Pietikäinen and D. Harwood, A comparative study of texture measures with classification based on featured distributions, *Pattern Recognition* **29**(1), pp. 51–59 (1996).

[47] C. E. Shannon, Communication in the presence of noise, in *Proceedings of the Institute of Radio Engineers* (IEEE) **37**(1), pp. 10–21 (1949).

[48] X. Tan and B. Triggs, Enhanced local texture feature sets for face recognition under difficult lighting conditions, *IEEE Transactions on Image Processing* **19**(6), pp. 1635–1650 (2010), doi:10.1109/TIP.2010.2042645.

[49] R. Gonzalez-Diaz, W. G. Kropatsch, M. Cerman and J. Lamar, Characterizing configurations of critical points through lbp, in D. M. Onchis and P. Real Jurado (eds.), *Proceedings of the 5th International Workshop on Computational Topology in Image Context, CTIC2014*, Timisoara, Romania (2014).

[50] M. Cerman, I. Janusch, R. Gonzalez-Diaz and W. G. Kropatsch, Topology-based image segmentation using LBP pyramids, *Machine Vision and Applications* **27**(8), pp. 1161–1174 (2016).

[51] L. Latecki, U. Eckhardt and A. Rosenfeld, Well-composed sets, *Computer Vision and Image Understanding* **61**(1), pp. 70–83 (1995).

[52] M. Cerman, R. Gonzalez-Diaz and W. G. Kropatsch, LBP and irregular graph pyramids, in N. Petkov and G. Azzopardi (eds.), *Computer Analysis of Images and Patterns*, Malta. Lecture Notes in Computer Science, Vol. 9256. Springer, Cham (2015).

[53] W. G. Kropatsch, R. M. Casablanca, D. Batavia and R. Gonzalez-Diaz, Computing and reducing slope complexes, in R. Marfil, M. Calderon, F. D. del Rio, P. Real and A. Banderas (eds.), *Proceedings 7th International Workshop on Computational Topology in Image Context*, Malaga, Spain. Lecture Notes in Computer Science, Vol. 11382. Springer, Berlin, pp. 12–25 (2019).

[54] D. R. Martin, C. Fowlkes, D. Tal and J. Malik, A database of human segmented natural images and its application to evaluating segmentation algorithms and measuring ecological statistics, in *8th International Conference on Computer Vision, ICCV 2001*, Vol. 2, pp. 416–423 (2001).

[55] D. Batavia, R. Gonzalez-Diaz and W. G. Kropatsch, Image = structure + few colors, in A. Torsello, L. Rossi, M. Pelillo, B. Biggio and A. Robles-Kelly (eds.), *S+SSPR 2020*. Lecture Notes in Computer Science, Vol. 12644. Springer Nature, pp. 365–376 (2021).

[56] Z. Wang, A. Bovik, H. Sheikh and E.P. Simoncelli, Image quality assessment: From error visibility to structural similarity, *IEEE Transactions on Image Processing* **13**(4), pp. 600–612 (2004).

[57] L. Zhang, L. Zhang, X. Mou and D. Zhang, FSIM: A feature similarity index for image quality assessment, *IEEE Transactions on Image Processing* **20**(8), pp. 2378–2386 (2011), doi:10.1109/TIP.2011.2109730.

[58] W. G. Kropatsch, R. M. Casablanca, D. Batavia and R. Gonzalez-Diaz, On the space between critical points, in M. Couprie, J. Cousty, Y. Kenmochi and N. Mustafa (eds.), *Discrete Geometry for Computer Imagery*, Marne-la-Vallée, France. Lecture Notes in Computer Science, Vol. 11414. Springer, Berlin, pp. 115–126 (2019).

[59] R. Gonzalez-Diaz, D. Batavia, R. M. Casablanca and W. G. Kropatsch, Characterizing slope regions, *Journal of Combinatorial Optimization* **44**, pp. 1–20 (2021).

[60] D. Batavia, J. Hladůvka and W. G. Kropatsch, Partitioning 2D images into prototypes of slope region, in M. Vento and G. Percannella (eds.), *Proceedings of 18th International Conference on Computer Analysis of Images and Patterns*, Salerno, Italy. Lecture Notes in Computer Science, Vol. 11678. Springer Nature Switzerland AG, pp. 363–374 (2019).

[61] N. Passat and Y. Kenmochi, A topological tree of shapes, in B. E. (ed.), *DGMM 2022*, Vol. 13493. Springer Nature, Switzerland, pp. 221–235 (2022).

[62] B. N. Delaunay, Sur la sphère vide, *Bulletin of Academy of Sciences of the USSR* **7**(6), pp. 793–800 (1934).

© 2025 World Scientific Publishing Company
https://doi.org/10.1142/9789811289125_0003

Chapter 3

A Sensor-Independent Multi-Modal Fusion Scheme for Human Activity Recognition

Anastasios Alexiadis[*], **Alexandros Nizamis**[†],
Dimitra Zotou[‡], **Dimitrios Giakoumis**[§],
Konstantinos Votis[¶], **and Dimitrios Tzovaras**[∥]

*Centre for Research and Technology, Hellas,
Information Technologies Institute (CERTH/ITI),
6km Charilaou-Thermi, Thessaloniki, Greece*
[*]*talex@iti.gr*
[†]*alnizami@iti.gr*
[‡]*dzotou@iti.gr*
[§]*dgiakoum@iti.gr*
[¶]*kvotis@iti.gr*
[∥]*Dimitrios.Tzovaras@iti.gr*

Abstract

The field of human activity recognition serves as the foundational basis for the development and implementation of ambient intelligence and assisted living applications. Multi-modal approaches for human activity recognition employ various sensors and *fuse* them to get more precise outcomes. In order for these methods to function effectively, it is necessary to have data available for all sensors that are being utilized. This chapter introduces a sensor-agnostic approach for developing multi-modal methods that can function effectively even in the absence of sensor data, regardless of the number of sensors employed. In addition, we propose a data augmentation method that enhances the accuracy of the fusion model by up to 11% in scenarios where sensor data are lacking. The efficacy of the suggested methodology is assessed using the ExtraSensory dataset, which consists of over 300,000 instances collected from 60 individuals using smartphones and smartwatches. Furthermore, the methodologies are assessed for varying quantities of concurrently employed sensors. However, it is necessary to have prior knowledge about the maximum number of sensors.

3.1 Introduction

The foundations of Ambient Intelligence (AmI) and Assisted Living Applications (AAL) can be traced back to the field of Automatic Human Activity Recognition (HAR). It involves the difficulties associated with identifying and comprehending human behaviors and their surrounding circumstances, which are essential prerequisites for incorporating computer decision-making abilities that are sensitive to human behavior. Human Activity Recognition (HAR) can be conducted by using static sensors, such as a mounted video camera [1], or wearable sensors, such as a smartwatch or other types of wearable sensors [2]. Alternatively, a combination of both static and wearable sensors can also be employed for the purposes of HAR.

There are several ways to accomplish HAR [3]. The two primary categories of human activity recognition technologies are unimodal and multi-modal, which are determined by the type of sensor utilized. Unimodal methods employ data derived from a solitary modality, such as an audio stream. The methods can be classified into four categories [3]: (i) space-time, (ii) stochastic, (iii) rule-based, and (iv) shape-based methods. In contrast, multi-modal approaches involve the integration of features derived from several sources, such as the combination of audio and visual sensors [4]. These methods can be classified into three categories: affective, behavioral, and social networking methods [3].

This study introduces a sensor-agnostic fusion technique that is not dependent on the quantity of sensors employed for HAR in smart homes. The term activity refers to the routine duties performed by individuals on a daily basis. Furthermore, we propose a data augmentation method that enhances the acquired data by incorporating subsets of sensor data in every observation. We next employ these methods to combine unimodal models for the ExtraSensory dataset [5]. The dataset was utilized as a test case for evaluating the efficacy of our proposed methodologies. The methods we employ are applicable to any model that integrates multiple unimodal models across a range of sensors. The ExtraSensory dataset comprises a substantial collection of more than 300,000 instances derived from a cohort of 60 individuals representing a wide range of ethnic origins. These instances were gathered via both smartphones and smartwatches. The dataset comprises diverse measurements obtained from a range of wearable sensors, such as accelerometers, gyroscopes, magnetometers, watch compasses, and audio sensors. Not all sensors were consistently accessible during the duration of the study. Certain phones lacked specific sensors, resulting in their unavailability. Similarly, there were instances where sensors were intermittently inaccessible.

The remaining chapters are organized as follows. The following section provides a quick overview of the associated literature. Thereafter, we discuss our approaches to data augmentation and sensor-independent (with respect to the number of sensors) fusion. On the basis of a series of experiments, we assess our methodology and review the outcomes. In the end, we conclude this chapter by outlining our goals for future work.

3.2 Related Work

In recent years, as a result of the proliferation of IoT devices and smart living environments, a great deal of research has been conducted on HAR, and numerous methods have been developed and implemented. The introductory section outlines the classification of approaches into five primary categories: space-time, stochastic, rule-based, shape-based, and multi-modal methods. This chapter aims to provide a concise literature review on multi-modal methods utilized for HAR. The purpose of this chapter is to position our work within the current state-of-the-art and highlight the innovative aspects and contributions of our research in this domain.

Several multi-modal methods for human activity recognition are based on fusion techniques [6] because an event or activity can be described by various categories of data and characteristics that provide additional and useful information. In [7], the multi-modal fusion for detecting human activity is further sub-divided into data and feature fusion methods. A decision fusion method is also considered in [23]. The latter is not a direct fusion scheme because it employs discrete classifiers to calculate probability scores and then combines them for decision-making [8, 9]. Since this chapter is considered as a more direct type of fusion, the relevant works in this section are presented based on data and feature methods' classification. The primary goal of data fusion in this context is to enhance the accuracy, robustness, and reliability of systems designed for recognizing human activities. This type of fusion involves the integration of data from various mobile and wearable sensor devices. The study conducted by [10] presents a deep learning framework called DeepSense, which aims to address the issues of noise and feature modification in order to enhance recognition accuracy. The framework leverages the interplay between several sensory modalities through the integration of convolutional and recurrent neural networks into sensor systems. In pursuit of the same objective, [11] put forth a deep learning framework that utilizes convolutional and LSTM recurrent units for the purpose of activity recognition. The framework exhibits compatibility with multi-modal wearable sensors, as it possesses the inherent capability to seamlessly integrate sensor fusion, hence obviating the need for specialized expertise in feature design. The authors in [12] provide a novel activity recognition system designed for the purpose of monitoring older individuals. The system gathers and integrates data from several sensors, including state sensors located at entrances, motion sensors installed in rooms, and sensors affixed to appliances or furniture items. However, the detection of activity is facilitated solely through the utilization of numerous Hidden Markov Models (HMMs). In [13], a novel approach has been proposed for human daily activity recognition, which involves the integration of HMMs with neural networks to facilitate multi-sensor fusion. The approach utilized wearable sensors and was evaluated within the context of robot-assisted living system designed for elderly people. Alternative methodologies such as [14] involve the amalgamation of extracted characteristics from experimental data obtained from various sensors, including a depth camera, an accelerometer, and a micro-Doppler radar. The authors utilize the data from the aforementioned sensors to generate combinations that are used for the purpose of classifying the activity. It was discovered that the incorporation of additional sensors resulted in a consistent enhancement of classification accuracy. The accuracy of the quadratic-kernel SVM classifier and the ensemble classifier has been measured by the authors.

Over the past years, researchers have created fusion methods at the features level with the aim of enhancing the accuracy and performance of activity detection systems. Feature fusion approaches facilitate the integration of features derived from sensor data with machine learning algorithms. This chapter also employs this form of fusion. The sensors utilized for activity recognition can be classified into several categories. First, 3D sensors have been employed to identify activities, such as walking, running, or sitting [15–17]. Second, thermal cameras have been utilized for recognizing household activities [18, 19]. Lastly, event cameras [20] have been employed for event-based activity detection and activity tracking applications [21]. In [22], a multi-modal feature-level fusion approach for robust human activity recognition was proposed. The methodology employs data collected from many sensors, including a depth sensor, an RGB camera, and wearable inertial sensors. The recognition framework underwent testing using a publicly accessible dataset comprising more than 25 distinct human activities. Support vector machine (SVM) classifiers and K-nearest neighbor algorithms were employed for the training and testing of the suggested fusion model. The researchers noted that employing a greater number of sensors in the fusion process yielded improved outcomes. However, the observed increase in accuracy was accompanied by a notable decrease (exceeding 10%) in performance when compared to fusion systems utilizing fewer sensor combinations. In another approach, [4], a method for human activity recognition that incorporates multi-modal feature selection and fusion methods using video data was presented. The researchers gathered audio and visual elements from a publicly available dataset or video and employed them as input for a series of SVM classifiers. The final classification score was obtained by fusing the outputs using a fuzzy integral and a two-layer SVM. The researchers noted that the utilization of audio context seems to be more advantageous compared to visual context. However, it is important to acknowledge that audio may not always be beneficial for some behaviors due to its extensive variability. Recently, in [23], the authors have put forth a novel approach to activity recognition in body sensor networks, which involves the integration of intelligent sensor fusion. This strategy is particularly relevant in scenarios where data are frequently ambiguous or partial. The methodology employed by the researchers was founded upon the principles of the Dezert–Smarandache theory. In this chapter, the researchers utilized Kernel Density Estimation (KDE)-based models to analyze the training dataset consisting of sensor readings. Subsequently, they identified the most effective discriminative model from the options available. The utilization of a testing dataset was also employed to compute fundamental belief assignments by employing KDE models for each activity. Ultimately, the belief assignments that were computed were amalgamated with redistribution criteria in order to facilitate the process of reaching final decisions. The researchers concluded that their methodology exhibited superior performance in terms of accuracy when compared to existing approaches that are considered to be at the forefront of the field. This conclusion was drawn based on the evaluation and comparison of their approach against other methods using two publicly available datasets.

The Tensor Deep Learning (TDL) [24] method proposes a high-order backpropagation algorithm that extends the data from the linear space to multiple linear space and subsequently trains the parameters of the proposed model. This method improves big data feature learning and high-level future fusion by using tensors to model the complexity of multi-source heterogeneous data and by extending the vector space data to tensor space when feature extraction in the tensor space is included.

When fusing information from different sensors, an essential part of the problem is synchronizing the information. When using a smartphone system clock for the synchronizing, timing errors ranging from milliseconds to multiple seconds were identified [25]. A classic approach used sliding-window mechanisms in IoT data streaming to divide the incoming large volume of data into manageable segments. A weighted adaptive partition algorithm which performs a dynamic weighted fusion of multi-source heterogeneous IoT data in sliding windows decreased the error rate of the fusion model [26].

Fusion has also been applied to medical imaging to diagnose diseases by combining different tissues of patients [27] as well as in Wearable IoT sensors for heartbeat detection [28]. Moreover, it has been applied for emotional classification prediction in AI Assisted telemedicine health analysis [29].

This chapter presents a novel approach to feature fusion that demonstrates superior accuracy and resilience in human activity recognition when compared to previously discussed data fusion techniques. Its novelties lie in the following points: The method described in this chapter offers a fusion mechanism that is not dependent on the number of sensors being used nor at its types. In contrast to the fusion methodologies discussed in the preceding paragraph, our methodology does not necessitate the reconstruction of the fusion model when fewer sensors are accessible. An further enhancement brought out by this method, in contrast to the fusion methods discussed above, is the utilization of a data augmentation strategy. This methodology enhances the compiled data by using subsets of sensor data that are utilized for each individual observation. By incorporating various combinations of existing sensor data, the volume of available data is augmented, hence enhancing the accuracy of the system in terms of human activity recognition when operating with fewer available sensor data.

3.3 Methods

For the purposes of this chapter, seven unimodal classifiers were provided, each of which categorizes everyday activities of humans in accordance with a particular sensor from the ExtraSensory dataset. These classifiers are regarded as black boxes. Each classifier applies to one of the following sensors:

- Watch Accelerometer (WA),
- Watch Compass (WC),
- Phone Accelerometer (PA),
- Phone Gyroscope (PG),
- Phone Magnet (PM),
- Phone State (PS),
- Audio (A).

In Table 3.1, the F1 scores of the seven classifiers with regards to their respective test sets (which are subsets of the test set containing only the observations that contain data for their respective sensors)

Table 3.1. Unimodal classifiers F1-scores.

Classes	WA F1	WC F1	PA F1	PG F1	PM F1	PS F1	A F1
Sitting-Toilet	0.23	0.0	0.0	0.0	0.0	0.08	0.13
Sitting-Eating	0.7	0.38	0.064	0.0	0.0	0.26	0.42
Standing-Cooking	0.09	0.0	0.0	0.13	0.13	0.0	0.33
Sitting-Watching TV	0.05	0.0	0.67	0.57	0.57	0.72	0.83
Lying down-Watching TV	0.63	0.0	0.28	0.18	0.18	0.37	0.58
Standing-Eating	0.24	0.49	0.47	0.22	0.22	0.62	0.69
Standing-Cleaning	0.63	0.0	0.4	0.19	0.19	0.69	0.77
Walking-Eating	0.43	0.0	0.27	0.38	0.38	0.43	0.6
Standing-Watching TV	0.33	0.46	0.3	0.27	0.27	0.54	0.59
Standing-Toilet	0.29	0.31	0.59	0.5	0.5	0.7	0.79
Walking-Watching TV	0.49	0.0	0.1	0.08	0.08	—	0.63
Walking-Cooking	0.42	0.0	0.39	0.11	0.11	0.37	0.51
Sitting-Cooking	0.59	0.0	0.35	0.38	0.38	0.67	0.68
Walking-Cleaning	0.05	0.0	0.19	0.1	0.1	0.11	0.48
Lying down-Eating	0.16	0.0	0.62	0.58	0.58	0.24	0.73
Sitting-Cleaning	0.15	0.0	0.0	0.0	0.0	0.16	0.27
Accuracy	0.55	0.4	0.5	0.45	0.45	0.65	0.74
Macro avg	0.34	0.1	0.29	0.23	0.23	0.4	0.57
Weighted avg	0.55	0.34	0.52	0.43	0.43	0.64	0.75

are presented. A dash "–" denotes that there were no observations of the respective sensor for that class. There are 16 daily activity classes, which are presented in Table 3.1. Each of the seven classifiers contains a 32-node dense layer followed by a softmax layer at their end.

3.3.1 *Flow of fusion and training process*

Figure 3.1 presents the flow of the training and fusion process. The data from the 7 sensors are utilized to train their corresponding unimodal models. Subsequently, the unimodal models are utilized to compute the feature vectors of their corresponding sensor data and these are passed to Algorithm #1 to create the fusion data set. This set can either be utilized to train the AI Fusion model or it can be augmented first by using Algorithm #2. Both of these algorithms are presented in the following section.

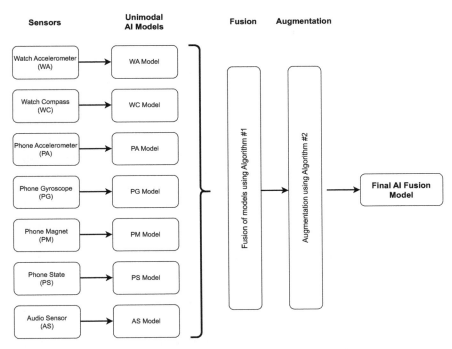

Fig. 3.1. Flow of the training and fusion process diagram.

3.3.2 Sensor independent fusion model

The fusion model employed a feed-forward artificial neural network. The softmax layer of each unimodal model was omitted during the simulation process, resulting in a feature vector of size 32, which corresponds to the output of the penultimate layer of the unimodal models. The feature vectors serve as inputs for the softmax layers of their corresponding unimodal models. These vectors contain the extracted activity information from the data, which is subsequently transformed into a probability distribution by the softmax layer. The simulation of each unimodal model is conducted by utilizing the design attributes derived from observations made by its corresponding sensor.

The input layer is defined with a length of 32 multiplied by the number of sensors. To provide sensor independence in respect to the number of sensors used, we add a binary vector to the input layer of the network with the length equal to the number of sensors. So, for our specific case, the input layer has a size of $32 \cdot 7 + 7 = 231$.

$$F_{s1_1} F_{s1_2} ... F_{s1_{32}} ... F_{s7_1} F_{s7_2} ... F_{s7_{32}} ... F_{\text{acts}7} F_{\text{acts}6} ... F_{\text{acts}1}$$
$$\uparrow \qquad\qquad \uparrow \qquad\qquad\qquad\qquad\qquad\qquad\qquad\qquad\qquad \uparrow$$

The diagram above illustrates the input layer of the fusion model. F_{s1_1} denotes the first feature of the first sensor and $F_{s1_{32}}$ denotes the last feature of the first sensor, whereas $F_{\text{acts}1}$ denotes the binary feature that actives/deactivates sensor 1 input for the fusion model. When $F_{\text{acts}1}$ is set to 0, the features corresponding to sensor 1 input, that is, the features from F_{s1_1} to $F_{s1_{32}}$, are set to 0 too. When $F_{\text{acts}1}$ is set to 1, the input features corresponding to sensor 1 on the input of the fusion model are set to the 32 values of the feature vector which is the output of the respective unimodal model for sensor 1. In a similar manner, we set the other sensor inputs. The fusion model performs feature-level fusion.

Algorithm 1 computes the dataset for training the fusion model. It is given the training set ($train$) containing the features for all sensors, a dictionary ($feature\_sensors$) with mappings of the form $sensor\_name \rightarrow [feature\_indeces]$, for each sensor providing the indeces for each sensor's features in the training set, a dictionary of the unimodal models ($unimodal\_models$) of the form

Algorithm 1: Create Fusion Model Training Set

1: **procedure** CREATE_DATASET($train, feature\_sensors, unimodal\_models, sensors$)
2:     $T \leftarrow [0]_{len(train) \times 32 \cdot len(sensors) + len(sensors)}$ ▷ Matrix of zeros
3:     **for** each sensor k ranging from 0 to $len(sensors) - 1$ **do**
4:        $train\_s \leftarrow train[feature\_sensors[sensors[k]]]$
5:        $train\_sn \leftarrow train\_s.dropna()$
6:        $idxs \leftarrow train\_sn.index$
7:        $feature\_matrix \leftarrow simulate\_model(unimodal\_models[sensors[k]], train\_sn)$
8:        **for** each observation i ranging from 0 to $len(T) - 1$ **do**
9:           **if** $i \in idxs$ **then**
10:              $T[i, T.no\_of\_cols - k - 1] \leftarrow 1$
11:           **end if**
12:        **end for**
13:        **for** each observation i ranging from 0 to $len(feature\_matrix) - 1$ **do**
14:           $T[idxs[i], k \cdot 32 : (k+1) \cdot 32] \leftarrow feature\_matrix[i]$
15:        **end for**
16:     **end for**
17:     **return** T
18: **end procedure**

$sensor\_name \rightarrow model$ and a list of the sensor names ($sensors$). For each observation in the training set, the feature vectors of the unimodal models are computed, in the cases when there are data available for their respective sensors, and the activating feature for these sensors is set to 1. When data are missing for a sensor, the respective features for that sensor are set to 0 as well as the activating feature.

3.3.3 Data augmentation method

A data augmentation method was devised and executed with the underlying assumption that the dataset can be enlarged by incorporating additional observations that encompass all conceivable subsets of activated sensors derived from the sensors carrying data in each original dataset observation. For instance, let us examine the scenario in which additional data are incorporated based on a solitary observation, wherein just five out of the total seven sensors are employed. The following combinations are available with only the five sensors activated:

In each of these instances, the target activity is known and shares the same label as the original observation, which encompasses data from all sensors. Consequently, we can augment the dataset by including a new observation for each case. This new observation will provide data for the relevant sensors and set the features of the missing sensor data, along with their corresponding activation features, to 0 in the training set of the fusion model.

```
[('s1', 's2', 's3', 's4', 's5'),
 ('s1', 's2', 's3', 's4', 's6'),
 ...,
 ('s2', 's4', 's5', 's6', 's7'),
 ('s3', 's4', 's5', 's6', 's7')]
```

The proposed method commences by setting the value k to equal the activated sensors of each observation and loops, decreasing the number of used sensors by 1 in each iteration and computes the combinations, that is the k-length tuples with no repetition for each value of k until $k < C$, where C is a constant defining the minimum number of sensors that can be utilized. No interpolation or estimation techniques are performed to augment the dataset; the labels for the generated data are already known as well as the sensor data used for the new observations are the feature vectors computed by the unimodal models for the measured sensor data.

Algorithm 2 enhances the dataset based on the above method. The dataset for the fusion model (*data*) is provided, which is obtained through the execution of Algorithm 1. Additionally, we have the the vector of labels (Y) that indicates the class for each observation, the number of sensors (*no_of_sensors*), and the constant C, which determines the minimum number of sensors required for the augmented dataset. For each observation in the training set, all $k - tuple$ combinations without repetition are computed, ranging from k equal to the number of activated sensors down to C.

A Sensor-Independent Multi-Modal Fusion Scheme

Algorithm 2: Data Augmentation Method

1: **procedure** AUGMENT_DATA($data, Y, no\_of\_sensors, C$)
2:     $aug\_data \leftarrow ()$ ▷ Empty Sequence
3:     $aug\_Y \leftarrow ()$ ▷ Empty Sequence
4:     $sensor\_ids \leftarrow \{z : \exists n \in \mathbb{Z} \text{ such that } z = 0 + 1 \times n, \text{ and } z \in [0, no\_of\_sensors - 1)\}$
5:     **for** each observation i ranging from 0 to $len(data) - 1$ **do**
6:         $sources\_used \leftarrow \sum_{x=data.no\_of\_cols - no\_of\_sensors}^{data.no\_of\_cols} data[i, x]$
7:         **for** each sensor k ranging from $no\_of\_sensors - 1$ to 0 in steps of -1 **do**
8:             **if** $k < sources\_used$ **then**
9:                 **if** $k < C$ **then**
10:                     **break**
11:             **end if**
12:             $v \leftarrow$
$$\binom{sensor\_ids}{k} \quad (3.1)$$
                ▷ k-length tuples with no repetition
13:             **for** l ranging from 0 to $len(v) - 1$ **do**
14:                 $OBV \leftarrow [0]_{data.no\_of\_cols}$ ▷ Vector of zeros
15:                 **for** m ranging from 0 to $len(v[l]) - 1$ **do**
16:                       $OBV[v[l][m] \cdot 32 : (v[l][m]+1) \cdot 32] \leftarrow data[i, v[l][m] \cdot 32 : (v[l][m] + 1) \cdot 32]$
17:                       $OBV[data.no\_of\_cols - v[l][m] - 1] \leftarrow 1$
18:                 **end for**
19:                 $aug\_data \leftarrow aug\_data \frown (OBV)$ ▷ Sequence Concatenation
20:                 $aug\_Y \leftarrow aug\_Y \frown (Y[i])$
21:             **end for**
22:             **else**
23:                 $aug\_data \leftarrow aug\_data \frown (data[i,:])$
24:                 $aug\_Y \leftarrow aug\_Y \frown (Y[i])$
25:             **end if**
26:         **end for**
27:     **end for**
28:     **return** $as\_matrix(aug\_data), as\_vector(aug\_Y)$
29: **end procedure**

3.4 Evaluation

In order to evaluate the proposed methods, a series of experiments was designed to examine the fusion model's improvement of F1 score per class. This evaluation was conducted on the entire test set as well as specific subsets of the test set that were divided based on the number of sensors utilized each observation. The dataset was partitioned into a training set and a test set, with a ratio of 70% to 30%, respectively. The train and test sets utilized for training and evaluating the unimodal models were derived from this process. During the training process, the initial training set was partitioned into two subsets, with a ratio of 80% for the final training set and 20% for the validation set. The fusion model was trained using the Adam optimizer with a learning rate of $lr = 0.001$. A $batch\_size = 64$ was utilized for 200 epochs, using early stopping with $patience = 50$ for validation accuracy, while reducing the learning rate when the validation accuracy has stopped improving using $factor = 0.1, patience = 2$. The training set was shuffled for the training process.

The outcomes of these experiments are presented in Table 3.2. The F1 columns represent the F1 scores per number of sensors used, whereas the S columns denote the corresponding samples provided in the test set (*support* metric), e.g., *3s F1* denotes the F1 scores for the subset holding the observations which utilized data from exactly three sensors, whereas *3s S* denotes the number of samples provided for that subset (*support*). *As* denotes that the whole test set was used (*all sensor data*). The resulting model provided an accuracy of 82% on the whole test set, whereas the subset containing the observations which hold data for all seven sensors provided an accuracy of 84%. According to the support metrics, there were fewer observations that provided data for less than five sensors. The classes with larger sample sizes (e.g., *Sitting-Watching TV* and *Standing-Toilet*) had accuracy scores of at least 86%. On the other hand, spectrum classes with too few samples (e.g., *Sitting-Toilet* and *Standing-Cooking*) had low scores of 38% and 55% respectively indicating balancing issues between the classes.

The next set of experiments investigates the proposed data augmentation's method performance. Table 3.3 presents the result of the experiments with the test set augmented with a value of $C = 2$ by using Algorithm 2. In Table 3.4, both the training/validation sets

Table 3.2. Fusion model with no data augmentation in training/validation sets and no data augmentation in test set. Xs F1 denotes F1 scores for X sensors and Xs S denotes support.

Classes	2s F1	2s S	3s F1	3s S	4s F1	4s S	5s F1	5s S	6s F1	6s S	7s F1	7s S	As F1	As S
Sitting-Toilet	0.50	2	0.50	2	0.84	22	0.82	145	0.69	64	0.67	18	0.38	17
Sitting-Eating	0.90	14	0.67	4	0.70	18	0.89	69	0.52	9	0.94	154	0.55	30
Standing-Cooking	0.67	2	1.00	2	0.51	17	0.00	1	0.71	104	0.71	12	0.55	14
Sitting-Watching TV	—	—	0.80	4	0.83	47	0.87	611	0.50	2	0.43	7	0.87	2294
Lying down-Watching TV	0.92	6	0.89	4	0.44	4	0.71	119	0.85	137	0.59	59	0.70	228
Standing-Eating	0.97	18	0.95	30	0.50	8	0.85	328	0.65	51	0.70	90	0.78	479
Standing-Cleaning	—	—	—	—	0.83	5	0.25	3	0.53	9	0.88	942	0.89	401
Walking-Eating	—	—	—	—	0.67	2	0.59	15	0.86	588	0.65	27	0.73	523
Standing-Watching TV	—	—	—	—	0.00	1	0.77	22	0.76	76	0.89	830	0.68	227
Standing-Toilet	0.00	1	1.00	2	0.88	109	0.72	61	0.83	340	0.72	234	0.86	1522
Walking-Watching TV	—	—	0.67	3	0.38	9	0.75	104	0.73	117	—	—	0.70	209
Walking-Cooking	—	—	—	—	0.00	1	0.84	26	0.50	3	0.78	208	0.72	80
Sitting-Cooking	—	—	—	—	—	—	0.50	1	0.81	36	0.73	60	0.74	59
Walking-Cleaning	—	—	—	—	0.67	3	0.67	13	0.76	29	0.87	44	0.52	57
Lying down-Eating	—	—	—	—	—	—	0.64	59	0.67	5	0.60	23	0.79	96
Sitting-Cleaning	—	—	—	—	—	—	—	—	0.51	35	0.46	8	0.58	9
Accuracy	0.88	43	0.88	51	0.77	246	0.82	1577	0.80	1605	0.84	2716	0.82	6245
Macro avg	0.66	43	0.81	51	0.56	246	0.66	1577	0.68	1605	0.71	2716	0.69	6245
Weighted avg	0.88	43	0.88	51	0.78	246	0.82	1577	0.80	1605	0.84	2716	0.82	6245

Table 3.3. Fusion model with no data augmentation in training/validation sets and with data augmentation in test set. Xs F1 denotes F1 scores for X sensors and Xs S denotes support.

Classes	2s F1	2s S	3s F1	3s S	4s F1	4s S	5s F1	5s S	6s F1	6s S	7s F1	7s S	As F1	As S
Sitting-Toilet	0.08	357	0.10	595	0.14	595	0.22	337	0.54	135	0.63	18	0.14	1949
Sitting-Eating	0.19	630	0.25	1050	0.34	1050	0.81	6180	0.90	1215	0.92	154	0.32	3453
Standing-Cooking	0.24	294	0.32	490	0.40	490	0.56	294	0.68	86	0.80	12	0.40	1666
Sitting-Watching TV	0.65	47894	0.73	78780	0.79	75044	0.82	32741	0.23	58	0.32	7	0.75	242583
Lying down-Watching TV	0.36	4748	0.46	7808	0.54	7533	0.59	3295	0.85	7182	0.60	59	0.49	24168
Standing-Eating	0.50	10059	0.59	16765	0.66	16017	0.68	6697	0.73	1560	0.71	90	0.62	51306
Standing-Cleaning	0.46	8040	0.61	13234	0.73	12618	0.47	570	0.67	694	0.88	942	0.64	41442
Walking-Eating	0.41	10943	0.50	18099	0.56	16497	0.62	7490	0.87	6150	0.68	27	0.52	55018
Standing-Watching TV	0.40	4626	0.49	7632	0.57	7494	0.55	2414	0.67	1755	0.90	830	0.51	22690
Standing-Toilet	0.57	31899	0.67	53029	0.75	52447	0.81	24898	0.58	464	0.71	234	0.70	169256
Walking-Watching TV	0.38	4327	0.48	7142	0.56	6834	0.64	2915	0.68	496	—	—	0.52	21776
Walking-Cooking	0.19	1680	0.28	2800	0.40	2732	0.51	1198	0.57	218	0.76	208	0.32	8655
Sitting-Cooking	0.41	1239	0.51	2065	0.59	1895	0.59	614	0.83	344	0.71	60	0.53	6002
Walking-Cleaning	0.21	1197	0.28	1995	0.35	1961	0.40	916	0.63	166	0.84	44	0.31	6168
Lying down-Eating	0.50	2016	0.59	3360	0.68	3326	0.78	1695	0.37	91	0.62	23	0.64	10785
Sitting-Cleaning	0.06	189	0.10	247	0.14	143	0.18	64	0.11	3	0.43	8	0.10	646
Accuracy	0.53	130138	0.63	215091	0.70	206676	0.76	92318	0.81	20617	0.84	2716	0.65	667563
Macro avg	0.35	130138	0.43	215091	0.51	206676	0.58	92318	0.62	20617	0.70	2716	0.47	667563
Weighted avg	0.54	130138	0.63	215091	0.70	206676	0.76	92318	0.81	20617	0.84	2716	0.66	667563

Table 3.4. Fusion model with data augmentation in training/validation sets and with data augmentation in test set. Xs F1 denotes F1 scores for X sensors and Xs S denotes support.

Classes	2s F1	2s S	3s F1	3s S	4s F1	4s S	5s F1	5s S	6s F1	6s S	7s F1	7s S	As F1	As S
Sitting-Toilet	0.13	357	0.19	595	0.23	595	0.31	337	0.60	135	0.63	18	0.21	1949
Sitting-Eating	0.31	630	0.39	1050	0.46	1050	0.86	6180	0.91	1215	0.93	154	0.43	3453
Standing-Cooking	0.44	294	0.53	490	0.61	490	0.71	294	0.80	86	0.86	12	0.57	1666
Sitting-Watching TV	0.72	47894	0.78	78780	0.82	75044	0.85	32741	0.29	58	0.44	7	0.79	242583
Lying down-Watching TV	0.45	4748	0.54	7808	0.60	7533	0.64	3295	0.87	7182	0.60	59	0.56	24168
Standing-Eating	0.59	10059	0.66	16765	0.69	16017	0.71	6697	0.74	1560	0.71	90	0.66	51306
Standing-Cleaning	0.65	8040	0.75	13234	0.81	12618	0.56	570	0.68	694	0.89	942	0.77	41442
Walking-Eating	0.48	10943	0.57	18099	0.62	16497	0.66	7490	0.89	6150	0.59	27	0.58	55018
Standing-Watching TV	0.45	4626	0.54	7632	0.61	7494	0.57	2414	0.69	1755	0.91	830	0.55	22690
Standing-Toilet	0.69	31899	0.76	53029	0.81	52447	0.85	24898	0.58	464	0.72	234	0.78	169256
Walking-Watching TV	0.49	4327	0.57	7142	0.63	6834	0.71	2915	0.73	496	—	—	0.60	21776
Walking-Cooking	0.38	1680	0.49	2800	0.57	2732	0.59	1198	0.58	218	0.76	208	0.51	8655
Sitting-Cooking	0.60	1239	0.67	2065	0.72	1895	0.68	614	0.86	344	0.76	60	0.67	6002
Walking-Cleaning	0.24	1197	0.29	1995	0.32	1961	0.32	916	0.72	166	0.91	44	0.29	6168
Lying down-Eating	0.66	2016	0.72	3360	0.76	3326	0.82	1695	0.31	91	0.76	23	0.74	10785
Sitting-Cleaning	0.35	189	0.41	247	0.39	143	0.35	64	0.21	3	0.33	8	0.38	646
Accuracy	0.64	130138	0.71	215091	0.76	206676	0.79	92318	0.83	20617	0.85	2716	0.72	667563
Macro avg	0.48	130138	0.55	215091	0.60	206676	0.64	92318	0.65	20617	0.72	2716	0.57	667563
Weighted avg	0.63	130138	0.71	215091	0.75	206676	0.79	92318	0.83	20617	0.85	2716	0.72	667563

Table 3.5. Fusion model with data augmentation in training/validation sets and no data augmentation in test set. Xs F1 denotes F1 scores for X sensors and Xs S denotes support.

Classes	2s F1	2s S	3s F1	3s S	4s F1	4s S	5s F1	5s S	6s F1	6s S	7s F1	7s S	As F1	As S
Sitting-Toilet	0.80	2	0.80	2	0.86	22	0.80	145	0.63	64	0.60	18	0.31	17
Sitting-Eating	0.86	14	0.67	4	0.73	18	0.83	69	0.60	9	0.93	154	0.52	30
Standing-Cooking	0.80	2	1.00	2	0.55	17	0.00	1	0.68	104	0.73	12	0.57	14
Sitting-Watching TV	—	—	0.89	4	0.76	47	0.87	611	0.44	2	0.50	7	0.88	2294
Lying down-Watching TV	1.00	6	0.75	4	0.43	4	0.67	119	0.85	137	0.63	59	0.66	228
Standing-Eating	0.97	18	0.95	30	0.43	8	0.84	328	0.64	51	0.68	90	0.76	479
Standing-Cleaning	—	—	—	—	0.67	5	0.00	3	0.33	9	0.89	942	0.88	401
Walking-Eating	—	—	—	—	0.57	2	0.56	15	0.87	588	0.62	27	0.70	523
Standing-Watching TV	—	—	—	—	0.00	1	0.78	22	0.74	76	0.91	830	0.72	227
Standing-Toilet	0.00	1	0.80	2	0.84	109	0.69	61	0.83	340	0.70	234	0.87	1522
Walking-Watching TV	—	—	0.67	3	0.29	9	0.82	104	0.71	117	—	—	0.68	209
Walking-Cooking	—	—	—	—	0.00	1	0.84	26	0.67	3	0.76	208	0.73	80
Sitting-Cooking	—	—	—	—	—	—	1.00	1	0.80	36	0.74	60	0.74	59
Walking-Cleaning	—	—	—	—	0.67	3	0.60	13	0.82	29	0.89	44	0.48	57
Lying down-Eating	—	—	—	—	—	—	0.60	59	0.55	5	0.69	23	0.80	96
Sitting-Cleaning	—	—	—	—	—	—	—	—	0.46	35	0.43	8	0.78	9
Accuracy	0.91	43	0.88	51	0.74	246	0.81	1577	0.80	1605	0.85	2716	0.82	6245
Macro avg	0.74	43	0.82	51	0.52	246	0.66	1577	0.66	1605	0.71	2716	0.69	6245
Weighted avg	0.90	43	0.88	51	0.74	246	0.81	1577	0.80	1605	0.84	2716	0.82	6245

are augmented with a value of $C = 2$ as well as the test set. Finally, in Table 3.5, only training and validation sets are augmented with a value of $C = 2$. As there are more combinations of 2-sensor data than 3-sensor data, and 3-sensor data than 4-sensor data, in Table 3.3, we can observe a sharp drop in the accuracy of the new augmented dataset, as it drops to 65%, whereas the accuracy of the subset containing data for all seven sensors remains at 84%. In Table 3.4, we observe an increase in total accuracy to 72%. When providing data for at least four sensors, the accuracy is higher, 76% for 4-sensor observations, 79% for 5-sensor observations, and 83% for 6-sensor observations. The largest increase in performance was 11% for 2-sensor observations. Note that according to Table 3.1 the sensor unimodal models do not perform equally. From this premise, it follows logically that not all sensor subsets will perform equally (especially the 2-sensor subsets). In Table 3.5, we can observe that the fusion model trained using the augmented training and validation sets performs equally to the non-augmented one on the original test set with an accuracy score of 82%. Its 2-sensor accuracy is higher though (91% vs. 88% when trained without the proposed data augmentation method).

3.5 Conclusions and Future Work

In this chapter, a novel fusion method is proposed for automatic human activity recognition that is independent of the number of sensors employed. The method employs feature-level fusion. This approach enables the development of fusion models capable of functioning with a reduced number of data sources compared to those for which the model was originally created. However, having prior knowledge about the maximum number of sensors is necessary.

In addition, we have presented a data augmentation method for enhancing the performance of the fusion model while dealing with observations containing limited sensor data. This method does not rely on interpolation or estimating techniques to fill the dataset. The algorithm generates all conceivable combinations of sensors employed for recorded observations, with a predetermined minimum threshold of sensor data necessary as specified by a constant value. The findings demonstrated a rise in all sub-divisions of the test dataset, categorized based on the quantity of sensors employed per observation, hence suggesting the efficacy of the approach.

In future research, we will explore class balancing methods as a means to enhance the performance of classes with limited sample sizes. Furthermore, we will also examine the specific contributions of each sensor to the outcomes, along with the disparities in accuracy observed when utilizing different combinations of sensors that yield the same total count. For instance, we will explore the disparity in accuracy when employing a 3-sensor dataset consisting of *Audio, Watch Compass, Phone Magnet* vs. *Phone State, Watch Accelerometer, Phone Accelerometer*. The objective is to develop a comprehensive approach for assessing the performance of a sensor within the fusion model, with the individual sensor's unimodal models being treated as opaque entities.

Acknowledgments

This research work was supported by the Hellenic Foundation for Research and Innovation (H.F.R.I.) under the "First Call for H.F.R.I. Research Projects to support Faculty members and Researchers and the procurement of high-cost research equipment grant" (Project Name: ACTIVE, Project Number: HFRI-FM17-2271).

References

[1] S. Zhang, Z. Wei, J. Nie, L. Huang, S. Wang and Z. Li, A review on human activity recognition using vision-based method, *Journal of Healthcare Engineering* **2017**, p. 3090343 (2017).

[2] M. Z. Uddin and A. Soylu, Human activity recognition using wearable sensors, discriminant analysis, and long short-term memory-based neural structured learning, *Scientific Reports* **11**(1), p. 16455 (2021).

[3] M. Vrigkas, C. Nikou and I. A. Kakadiaris, A review of human activity recognition methods, *Frontiers in Robotics and AI* **2**, p. 28 (2015), doi:10.3389/frobt.2015.00028.

[4] Q. Wu, Z. Wang, F. Deng, Z. Chi and D. D. Feng, Realistic human action recognition with multimodal feature selection and fusion, *IEEE Transactions on Systems, Man, and Cybernetics: Systems* **43**(4), pp. 875–885 (2013), doi:10.1109/TSMCA.2012.2226575.

[5] Y. Vaizman, K. Ellis and G. Lanckriet, Recognizing detailed human context in the wild from smartphones and smartwatches, *IEEE Pervasive Computing* **16**(4), pp. 62–74 (2017), doi:10.1109/MPRV.2017.3971131.

[6] B. Chandrasekaran, S. Gangadhar and J. M. Conrad, A survey of multisensor fusion techniques, architectures and methodologies, in *SoutheastCon 2017*. IEEE, pp. 1–8 (2017).

[7] H. F. Nweke, Y. W. Teh, G. Mujtaba and M. A. Al-Garadi, Data fusion and multiple classifier systems for human activity detection and health monitoring: Review and open research directions, *Information Fusion* **46**, pp. 147–170 (2019).

[8] M. Grabisch and E. Raufaste, An empirical study of statistical properties of the choquet and sugeno integrals, *IEEE Transactions on Fuzzy Systems* **16**(4), pp. 839–850 (2008).

[9] Z. Zeng, Z. Zhang, B. Pianfetti, J. Tu and T. S. Huang, Audio-visual affect recognition in activation-evaluation space, in *2005 IEEE International Conference on Multimedia and Expo*. IEEE, 4 p. (2005).

[10] S. Yao, S. Hu, Y. Zhao, A. Zhang and T. Abdelzaher, Deepsense: A unified deep learning framework for time-series mobile sensing data processing, in *Proceedings of the 26th International Conference on World Wide Web*, pp. 351–360 (2017).

[11] F. J. Ordóñez and D. Roggen, Deep convolutional and lstm recurrent neural networks for multimodal wearable activity recognition, *Sensors* **16**(1), p. 115 (2016).

[12] G. Sebestyen, I. Stoica and A. Hangan, Human activity recognition and monitoring for elderly people, in *2016 IEEE 12th International Conference on Intelligent Computer Communication and Processing (ICCP)*. IEEE, pp. 341–347 (2016).

[13] C. Zhu and W. Sheng, Multi-sensor fusion for human daily activity recognition in robot-assisted living, in *Proceedings of the 4th ACM/IEEE International Conference on Human Robot Interaction*, pp. 303–304 (2009).

[14] H. Li, A. Shrestha, F. Fioranelli, J. Le Kernec, H. Heidari, M. Pepa, E. Cippitelli, E. Gambi and S. Spinsante, Multisensor data fusion for human activities classification and fall detection, in *2017 IEEE SENSORS*. IEEE, pp. 1–3 (2017).

[15] Y.-S. Lee and S.-B. Cho, Activity recognition using hierarchical hidden Markov models on a smartphone with 3d accelerometer, in *International Conference on Hybrid Artificial Intelligence Systems*. Springer, pp. 460–467 (2011).

[16] J. K. Aggarwal and L. Xia, Human activity recognition from 3d data: A review, *Pattern Recognition Letters* **48**, pp. 70–80 (2014).

[17] L. Wang, D. Q. Huynh and P. Koniusz, A comparative review of recent kinect-based action recognition algorithms, *IEEE Transactions on Image Processing* **29**, pp. 15–28 (2019).

[18] G. Batchuluun, D. T. Nguyen, T. D. Pham, C. Park and K. R. Park, Action recognition from thermal videos, *IEEE Access* **7**, pp. 103893–103917 (2019).

[19] K. Naik, T. Pandit, N. Naik and P. Shah, Activity recognition in residential spaces with internet of things devices and thermal imaging, *Sensors* **21**(3), p. 988 (2021).

[20] S. U. Innocenti, F. Becattini, F. Pernici and A. Del Bimbo, Temporal binary representation for event-based action recognition, in *2020 25th International Conference on Pattern Recognition (ICPR)*. IEEE, pp. 10426–10432 (2021).

[21] J. Barrios-Avilés, T. Iakymchuk, J. Samaniego, L. D. Medus and A. Rosado-Muñoz, Movement detection with event-based cameras: Comparison with frame-based cameras in robot object tracking using powerlink communication, *Electronics* **7**(11), p. 304 (2018).

[22] M. Ehatisham-Ul-Haq, A. Javed, M. A. Azam, H. M. Malik, A. Irtaza, I. H. Lee and M. T. Mahmood, Robust human activity recognition using multimodal feature-level fusion, *IEEE Access* **7**, pp. 60736–60751 (2019).

[23] Y. Dong, X. Li, J. Dezert, M. O. Khyam, M. Noor-A-Rahim and S. S. Ge, Dezert-smarandache theory-based fusion for human activity recognition in body sensor networks, *IEEE Transactions on Industrial Informatics* **16**(11), pp. 7138–7149 (2020).

[24] W. Wang and M. Zhang, Tensor deep learning model for heterogeneous data fusion in internet of things, *IEEE Transactions on Emerging Topics in Computational Intelligence* **4**(1), pp. 32–41 (2020), doi:10.1109/TETCI.2018.2876568.

[25] S. S. Sandha, J. Noor, F. M. Anwar and M. Srivastava, Time awareness in deep learning-based multimodal fusion across smartphone platforms, in *2020 IEEE/ACM 5th International Conference on Internet-of-Things Design and Implementation (IoTDI)*, pp. 149–156 (2020), doi:10.1109/IoTDI49375.2020.00022.

[26] M. Zhang and W. Wang, Weighted adaptive partition for heterogeneous iot data stream, *IEEE Internet of Things Journal* **8**(20), pp. 15240–15248 (2021), doi:10.1109/JIOT.2020.3045726.

[27] Z. Qu, H. Jing, G. Bai, Z. Gao, L. Yu, Z. Zhang, G. Zhai and C. Yang, Computed tomography and 3d face scan fusion for IoT-based diagnostic solutions, *IEEE Internet of Things Journal*, pp. 1–1 (2023), doi:10.1109/JIOT.2023.3244201.

[28] A. John, S. J. Redmond, B. Cardiff and D. John, A multimodal data fusion technique for heartbeat detection in wearable iot sensors, *IEEE Internet of Things Journal* **9**(3), pp. 2071–2082 (2022), doi:10.1109/JIOT.2021.3093112.

[29] H. Yu and Z. Zhou, Optimization of iot-based artificial intelligence assisted telemedicine health analysis system, *IEEE Access* **9**, pp. 85034–85048 (2021), doi:10.1109/ACCESS.2021.3088262.

© 2025 World Scientific Publishing Company
https://doi.org/10.1142/9789811289125_0004

Chapter 4

Metrics for Saliency Map Evaluation of Deep Learning Explanation Methods

Tristan Gomez[*,‡], Thomas Fréour[†,§],
and Harold Mouchère[*,¶]

[*]*Nantes Université, Centrale Nantes,
CNRS, LS2N, Nantes, France*
[†]*Nantes University Hospital, Inserm,
CRTI, Inserm UMR 1064, Nantes, France*
[‡]*tristan.gomez@univ-nantes.fr*
[§]*thomas.freour@chu-nantes.fr*
[¶]*harold.mouchere@univ-nantes.fr*

Abstract

The black-box nature of deep learning models has spurred the development of visual explanation solutions for CNNs. However, conducting user studies to evaluate these methods is costly, necessitating the use of objective metrics for benchmarking visual explanations. In this chapter, we critically analyze the Deletion Area Under Curve (DAUC) and Insertion Area Under Curve (IAUC) metrics, designed to assess the faithfulness of saliency maps generated by methods like Grad-CAM or RISE. We reveal that these metrics overlook the actual saliency score values, focusing solely on score ranking, rendering them insufficient as the visual appearance of saliency maps can change significantly without altering the ranking. Additionally, we show quantitatively and qualitatively that during DAUC/IAUC computation, the model encounters out-of-distribution images, leading to unexpected model behavior. To address these concerns, we propose new metrics quantifying sparsity and calibration, previously unexplored properties. We also discuss related works evaluating the reliability of faithfulness metrics and highlight shared concerns about the metrics' inconsistent and unreliable nature due to out-of-distribution samples. Furthermore, we consider modifications to training and evaluation processes as potential solutions for the out-of-distribution issues. Finally, we provide general remarks on the metrics and discuss their potential evaluation in user studies.

4.1 Introduction

In recent years, interpretable machine learning has gained significant attention due to the lack of interpretability in deep learning models, which are widely used but often considered black boxes. Particularly, in image classification, various generic approaches have been proposed to generate saliency maps, highlighting salient areas of an image for the given task. However, evaluating these methods remains challenging due to the ambiguity of interpretability concept and the diverse user requirements in different applications, making it hard to establish a universal evaluation protocol.

Given the limitations of user studies in assessing interpretability and the associated high costs, another trend has emerged, focusing on designing objective metrics for evaluating generic explanation methods. Following this trend, we propose three new metrics and concentrate on studying the DAUC and IAUC metrics proposed in [1] with respect to two aspects. We demonstrate that these metrics overlook the actual saliency score values, relying solely on score ranking, rendering them inadequate as the visual appearance of saliency maps can change significantly without affecting the ranking. Additionally, we argue that the computation of DAUC and IAUC involves presenting the model with out-of-distribution images, potentially leading to unexpected model behavior and saliency map generation.

To address these issues, we introduce the sparsity metric, quantifying the sparsity of a saliency map, a previously neglected property. Furthermore, we propose two more metrics, Deletion Correlation (DC) and Insertion Correlation (IC), to measure the calibration of saliency maps. In conclusion, we provide general remarks on all the metrics examined in this chapter and discuss their potential evaluation in user studies. By focusing on objective metrics, we aim to advance the evaluation of generic explanation methods, ultimately contributing to more reliable and interpretable deep learning models.

4.2 Existing Metrics

Various metrics have been proposed to automatically evaluate saliency maps generated by explanation methods [1–3]. These metrics involve adding or removing important areas based on the saliency map and measuring the resulting impact on the initially predicted class score. For instance, Chattopadhay et al. introduced "Increase In Confidence" (IIC) and "Average Drop" (AD) metrics [3], which entail multiplying the input image with an explanation map to mask the non-relevant areas and then measuring the variation in the class score. Jung et al. proposed a variant of AD, called "Average Drop in Deletion" (ADD) [2], where the salient areas are masked instead of the non-salient ones. In parallel, Petsiuk et al. proposed DAUC and IAUC, which study the score variation while progressively masking/ revealing the image instead of applying the saliency map once [1]. Due to the conceptual similarities of all the mentioned metrics, our study focuses on DAUC and IAUC, which we now describe.

4.2.1 *DAUC and IAUC*

To evaluate the reliability of the proposed attention mechanism, Petsiuk et al. proposed the Deletion Area Under Curve (DAUC) and Integration Area Under Curve (IAUC) metrics [1]. These metrics evaluate the reliability of the saliency maps by progressively masking/revealing the image starting with the most important areas according to the saliency map and finishing with the least important.

The input image is a 3D tensor $I \in \mathbb{R}^{H \times W \times 3}$ and the saliency map is a 2D matrix $S \in \mathbb{R}^{H' \times W'}$ with a lower resolution, $H' < H$ and $W' < W$. First, S is sorted and parsed from the highest element to its lowest element. At each element $S_{i'j'}$, we mask the corresponding area of I by multiplying it by a mask $M^k \in \mathbb{R}^{H \times W}$, where

$$M^k_{ij} = \begin{cases} 0, & \text{if } i'r < i < (i'+1)r \text{ and } j'r < j < (j'+1)r \\ 1, & \text{otherwise,} \end{cases} \quad (4.1)$$

where $r = H/H' = W/W'$. After each masking operation, the model m runs an inference with the updated version of I, and the score of the initially predicted class is updated, producing a new score c_k:

$$c_k = m \left(I \cdot \prod_{\tilde{k}=1}^{\tilde{k}=k} M^{\tilde{k}} \right), \tag{4.2}$$

where $k \in \{1, \ldots, H' \times W'\}$. Examples of input images obtained during this operation can be seen in Figure 4.1. Second, once the whole image has been masked, the scores c_k are normalized by dividing them by the maximum $\max_k c_k$ and then plotted as a function of the proportion p_k of the image that is masked. The DAUC is finally obtained by computing the area under the curve (AUC) of this function. The intuition behind this is that if a saliency map highlights the areas that are relevant to the decision, masking them will result in a large decrease in the initially predicted class score, which in turn will minimize the AUC. Therefore, minimizing this metric corresponds to an improvement.

Instead of progressively masking the image, the IAUC metric starts from a blurred image and then progressively unblurs it by starting from the most important areas according to the saliency map. Similarly, if the areas highlighted by the map are relevant for predicting the correct category, the score of the corresponding class (obtained using the partially unblurred image) is supposed to increase rapidly. Conversely, maximizing this metric corresponds to an improvement.

4.2.2 Limitations

DAUC and IAUC generate out-of-distribution (OOD) images: When progressively masking/unblurring the input image, the model is presented with samples that can be considered out of the training distribution, as shown in Figure 4.1.

The distortions produced by masking/blurring operations are not naturally present in the dataset and differ from standard data augmentations, like random crop, horizontal flip, and color jitter. As a result, the model has not learned to process images with such distortions, leading to a distribution of presented images that is different

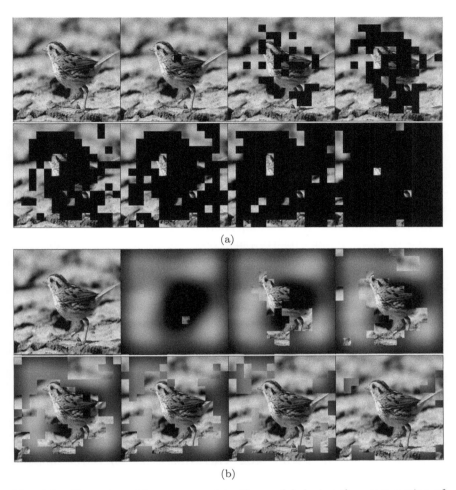

Fig. 4.1. Examples of images passed to the model during the computation of (a) DAUC and (b) IAUC. Masking and blurring the input images probably lead to OOD samples.

from what was encountered during training. CNNs and deep learning models, in general, are known to have poor generalization outside of the training distribution, as documented in previous studies [4]. Consequently, DAUC and IAUC may not effectively reflect the faithfulness of explanation methods, as they rely on model behavior that is different from what occurs during the test phase on the training distribution.

To test this hypothesis, we visualize the UMAP projections of representations for 100 masked/blurred samples obtained during the computation of DAUC and IAUC on the CUB-200-2011 dataset [5]. We also include the representation of 500 unmodified test images (in blue) to visualize the training distribution. Using a ResNet50 model [6] with Grad-CAM++ explanations [3], Figure 4.2 demonstrates that during the computation of DAUC, representations gradually converge to a single point, as all images are eventually fully masked (plain black). Even when only 40% of the image is masked, the corresponding representation remains distant from the blue point cloud representing the training distribution. A similar phenomenon occurs with IAUC, where blurring the image causes the representation to deviate from the training distribution.

To conduct a quantitative analysis, we propose evaluating the separability of image representations generated by the metrics from unmodified image representations in the test set. To achieve this, we train multiple models to distinguish representations from masked/blurred images (labeled as 1) from representations from unmodified images (labeled as 0). If any of these models demonstrate significant performance, it indicates that the two groups of images are separable, confirming the presence of an out-of-distribution (OOD) issue with blurred or black patches.

We train Support Vector Machine (SVM), K-Nearest Neighbors (KNN) classifier, Decision Tree (DT), and Multi-Layer Perceptron (MLP) models to classify representations obtained from ResNet-50 explained by Grad-CAM++ producing 14×14 resolution saliency maps, as shown in Figure 4.2. We utilize the same 100 images to form the dataset, consisting of representations from unmodified images and 196 representations of modified images with 1 to 196 black/blurred patches each (a total of 19600 images in class 1 and 100 images in class 0). Due to the highly unbalanced dataset, we measure model performance using the area under the ROC curve on the test set, with 40%, 10%, and 50% of the dataset used for training, validation, and testing, respectively. The evaluation is conducted using the `scikit-learn` library (version 1.0.2) with default parameter values. Results are presented in Table 4.1, showing that some models achieve good performances, indicating a distinct distribution of modified image representations compared to unmodified image representations. Notably, the performances are lower with blurred

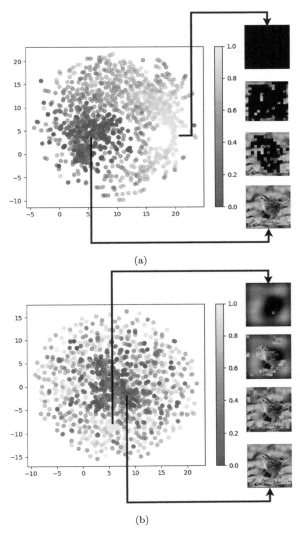

Fig. 4.2. UMAP projection of representations obtained while computing (a) DAUC and (b) IAUC on 100 images. The color indicates the proportion of the image that is masked/unblurred. The model used is a ResNet50 on which we applied Grad-CAM++ on the CUB-200-2011 dataset. We also plotted representations from 500 points of the test set to visualize the training distribution (in blue). By gradually masking the image, the representations converge toward a point (in yellow) that is distant from the points corresponding to unmasked images (in blue). Similarly, blurring the image causes the representation to move away from the training distribution. This shows that masking/blurring indeed creates OOD samples.

Table 4.1. Area under the ROC curve of models trained to distinguish image representations containing black/blurred patches from unmodified image representations. Values computed on the test set.

Model	Detection of	
	Black patches	Blurred patches
SVM	0.933	0.858
KNN	0.804	0.604
DT	0.814	0.711
NN	0.984	0.156

patches (IAUC column) than with black patches (DAUC column), suggesting that IAUC samples are less out of distribution than DAUC samples, which consistent with the observations in Figure 4.2.

In conclusion, our experiments confirm that both the DAUC and IAUC metrics produce out-of-distribution (OOD) samples, potentially leading to unexpected behavior of the model and the explanation map generation method. Interestingly, the blurring operation appears to create samples that are less distant from the training distribution compared to the masking operation, possibly because a blurred image retains low-frequency information from the original image, as suspected by Petsiuk et al. [1]. Additionally, it is worth noting that many current classification models assume the presence of an object to recognize in the input image, which contradicts DAUC and IAUC, as they involve removing the object of interest. Hence, modifying these metrics to always include an object for recognition in the input image could potentially resolve this issue. Further research and refinement of these metrics are necessary to enhance their performance and reliability in evaluating the faithfulness of explanation methods.

DAUC and IAUC only take the pixel score rank into account: During the computation of DAUC and IAUC, the saliency map is solely used to determine the order of masking/revealing the input image. Consequently, only the ranking of the saliency scores S_{ij} is considered to decide the masking sequence, disregarding the actual

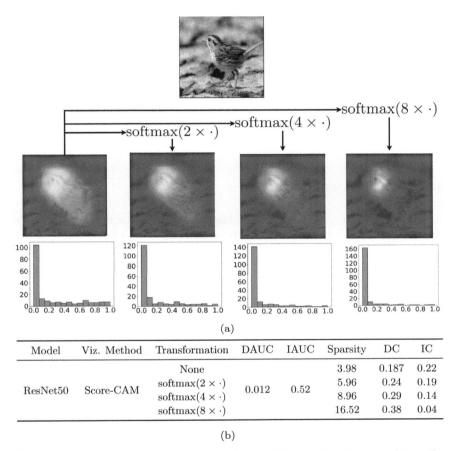

Fig. 4.3. Examples of saliency maps obtained by artificially sparsifying the saliency scores. The original saliency map is generated using Score-CAM applied on a ResNet50 model tested on the CUB-200-2011 dataset. Despite having different visual appearances, the four maps have the same DAUC and IAUC metric values because these metrics ignore the score values and only take into account the ranking of the scores. On the other hand, the sparsity metric depends on the score distribution and reflects the amount of focus of the map. In this figure, only the saliency map is modified, and the decision process is left unchanged.

score values. However, focusing solely on pixel ranking overlooks other crucial aspects, as two attention maps can exhibit significant visual variations without altering the ranking. As depicted in Figure 4.3, we demonstrate examples of a saliency map with artificially modified score distributions using Score-CAM [7] method. By adjusting the score distribution through a coefficient and the softmax function, we alter the visual appearance of the maps while maintaining the same DAUC and IAUC scores due to the unchanged pixel ranking. This observation highlights the limitation of DAUC and IAUC metrics in capturing the score dynamic of the saliency map, which substantially affects its visual appearance. To address this limitation, we propose new metrics in the following section that consider the score values to provide a more comprehensive evaluation of explanation methods.

4.3 Score Aware Metrics

As mentioned in the previous paragraph, DAUC and IAUC ignore the actual score values and only take into account the saliency score S_{ij} ranking. To complement these metrics, we propose three new metrics, namely sparsity, Deletion Correlation (DC), and Insertion Correlation (IC).

4.3.1 *The sparsity metric*

An important visual aspect of saliency maps that has not been studied until now by the community is what we call sparsity. As shown in Figure 4.3, saliency maps can be more or less focused on a specific point depending on the score distribution, without changing the score ranking. This aspect could impact the interpretability of the method, as it significantly changes the visual aspect of the map and therefore could also affect the perception of the user. For example, one could argue that a high sparsity value implies a map with a precise focus that highlights only a few elements of the input image, making it easier to understand by humans. The sparsity metric is defined as follows:

$$\text{Sparsity} = \frac{S_{\max}}{S_{\text{mean}}}, \qquad (4.3)$$

where $S_{\max}$ and S_{mean} are respectively the maximum and the mean score of the saliency map S. Note that the saliency methods available in the literature generate saliency maps with scores that are comprised of various types of ranges. Therefore, the map should be first normalized as follows:

$$S' = \frac{S - S_{\min}}{S_{\max} - S_{\min}}. \tag{4.4}$$

This means that, after normalization, $S'_{\max} = 1$ and Eq. (4.3) can be simplified to

$$\text{Sparsity} = \frac{1}{S'_{\text{mean}}} \tag{4.5}$$

A high sparsity value means a high $S_{\max}/S_{\text{mean}}$ ratio, i.e., a low mean score S_{mean} which indicates that the map's activated areas are narrow and focused. As shown in Figure 4.3(b), this metric is indeed sensitive to the actual saliency score values and reflects the various amount of focus observed in the saliency maps.

4.3.2 The DC and IC metrics

As mentioned earlier, the DAUC and IAUC metrics solely consider the ranking of saliency scores and disregard the actual score values. Consequently, these metrics overlook the sparsity of the saliency map, leading us to propose a method to quantify this aspect. Another noteworthy property of saliency maps is calibration, which has garnered recent attention in the deep learning community for calibrating prediction scores [8–10]. However, previous work has principally focused on calibrating prediction scores, leaving the calibration of explanatory maps understudied.

An ideally calibrated saliency map S would represent the importance of each pixel S_{ij} through its luminosity, reflecting its impact on the class score accurately. In other words, a perfectly calibrated explanatory map would satisfy the condition $S_{ij}/S_{i'j'} = v/v'$ for any two elements S_{ij} and $S_{i'j'}$, where v and v' represent the respective impacts of S_{ij} and $S_{i'j'}$ on the class score. To evaluate this calibration property, we propose quantifying the correlation between saliency scores and their corresponding impact on the class score. In practical applications, such a metric could be utilized in a user study to assess the usefulness and effectiveness of the calibration property.

Drawing inspiration from the DAUC and IAUC metrics, we introduce a new approach that involves progressively masking/revealing the input image based on the saliency map's order. However, instead of computing the area under the class score vs. pixel rank curve, we calculate the linear correlation between the variations in the class scores and the corresponding saliency scores. We refer to the correlation obtained during image masking as Deletion Correlation (DC) and the correlation during image revealing as Insertion Correlation (IC). The subsequent paragraph provides a detailed explanation of how these two metrics are computed.

For the computation of DC, we adopt the same progressive masking and inference method as in DAUC. After obtaining the class scores c_k, we calculate the score variations $v_k = c_k - c_{k+1}$ and then determine the linear correlation between the v_k and the corresponding saliency scores s_k, where s_k represents the saliency score of the area masked at step k.

Regarding the IC metric, we draw inspiration from IAUC. The process is initialized with a blurred image and gradually reveal the image following the saliency map's order. Once the image is entirely revealed, we calculate the score variations as $v_k = c_{k+1} - c_k$, and subsequently, we compute the linear correlation between the v_k and the corresponding saliency scores s_k. It's essential to note that the order of subtraction is reversed compared to DC because, during image revealing, we anticipate an increase in the class score.

In the computation of DC/IC for a well-calibrated saliency method, we anticipate a positive correlation between the class score variation and the saliency score. Specifically, when the class score variation is high, we expect the saliency score to be proportionally high as well, and conversely, when the class score variation is low, the saliency score should also be proportionally low.

The DC and IC metrics serve to measure calibration, a significant aspect overlooked by the DAUC, IAUC, and sparsity metrics. To illustrate this, we compute the DC and IC metrics for the examples depicted in Figure 4.3.

4.3.3 *Limitations*

The sparsity metric does not take into account the prediction scores: Indeed, this metric only considers the saliency score dynamic and ignores the class score produced by the model. However, this is not necessarily a problem as this metric was designed to be used as a complement to other metrics like DAUC, IAUC, DC, or IC, which takes the class score into account.

The DC and IC metrics also generate OOD images: As we took inspiration from DAUC and IAUC and also passed masked/blurred examples to the model, one can make the same argument as for DAUC/IAUC to show that the reliability of DC and IC could probably be improved by preventing OOD samples.

4.4 Benchmark

We compute the five metrics studied in the work (DAUC, IAUC, DC, IC, and sparsity) on *post hoc* generic explanation methods and attention architectures that integrate the computation of the saliency map in their forward pass. The *post hoc* methods are Grad-CAM [11], Grad-CAM++ [3], RISE [1], Score-CAM [7], Ablation CAM [12], and the activation map (AM) baseline. This baseline consists of simply visualizing the Euclidean norm of the feature vector of the last layer. The architectures with native attention are B-CNN [13], BR-NPA [14], the model from [15] which we call IBP (short for Interpretability By Parts), ProtoPNet [16], and ProtoTree [17]. Note that these attention models generate several saliency maps (or *attention* maps) per input image but the metrics are designed for a single saliency map per image. To compute the metrics on these models, we selected the first attention map among all the ones produced, as, in these architectures, the first is the most important one.

Training details: Models were trained using Pytorch [18] on two P100 GPUs. Images were resized to 448 × 448, pre-processed with standard ImageNet normalization, and augmented using color jittering, random crop, and horizontal flipping. Training is limited to 50 epochs, after which the best model is restored and used for final testing. The model used to generate the *post hoc* saliency maps is a regular ResNet-50 [6] model pre-trained on ImageNet [19]. The other models also use the same ResNet-50 as their backbone. As we were not able to retrain the ProtoPNet and ProtoTree models from scratch, we used the pre-trained weights proposed by the authors. The BR-NPA model is trained with the procedure proposed in the original paper [14]. First, a BR-NPA model is trained with an unmodified backbone which generates low-resolution attention maps (14 × 14). Then, we construct another BR-NPA model and modify the strides of layers 3 and 4 of the backbone to 1 to increase the final attention map size to 56 × 56. This higher-resolution model is then trained by distilling the first low-resolution model, as detailed in [14].

Hyperparameters: We optimized the following hyperparameters using the `optuna` [20] package: learning rate, training batch size, weight decay, and optimizer choice (among Adam, AMSGrad, and SGD). If SGD was chosen, we also optimized momentum and the choice of whether or not to use a scheduler. Finally, we also optimized the brightness and saturation parameters of the color jittering function and the crop ratio of the resizing function.

Resolution and metric computation: The *post hoc* methods and attention models used here generate saliency maps at various resolutions. The B-CNN, ProtoPNet, and ProtoTree models and the *post hoc* methods (excepted RISE) generate 14 × 14 saliency maps. The RISE method and the BR-NPA and IBP models respectively generate 7 × 7, 56 × 56, and 28 × 28 saliency maps. The number of inferences required for computing the faithfulness metrics is equal to the number of elements in the saliency map. As a consequence, the metric computation scale poorly with the saliency map resolution. For example, computing the DAUC metric on a single image with BR-NPA requires $56 \times 56 = 3136$ inferences. To limit computation for high-resolution models, several image regions are masked/unblurred before running an inference step. More precisely, we limit the number of total inferences to $14 \times 14 = 196$ and determine the number of areas modified per inference according to the following resolution: $3136/196 = 16$ for BR-NPA and $784/196 = 4$ for IBP.

During the computation of IAUC and IC, the blurred images are obtained by applying a 121 × 121 mean kernel, i.e., a kernel with all values set to $1/121^2$ with zero padding to maintain input size.

Table 4.2 shows the mean performances obtained on 100 random images from the CUB-200-2011 test set. The most important element to notice is the overall low values of correlation, especially

Table 4.2. Mean performance of explanations according to faithfulness metrics on 100 images of the CUB-200-2011 test set. Top half and bottom half respectively show performance with *post hoc* methods applied to ResNet-50 models and with attention modules added to the ResNet-50 backbones. All metrics have to be maximized to reflect an improvement except DAUC which should be minimized. Best performances are highlighted in bold.

Approach	Method	Accuracy ↑	DAUC ↓	IAUC ↑	DC ↑	IC ↑	Sparsity ↑
Post hoc	Ablation CAM		0.0215	0.26	0.36	−0.04	8.54
	Grad-CAM		0.0286	0.16	0.35	−0.12	5.28
	Grad-CAM++	0.842	0.0161	0.21	0.35	−0.07	6.73
	RISE		0.0279	0.18	**0.57**	−0.11	6.63
	Score-CAM		0.0207	0.27	0.32	−0.05	5.96
	AM		0.0362	0.22	0.31	−0.09	4.04
Attention model	B-CNN	0.848	0.0208	0.3	0.27	−0.02	12.74
	BR-NP	**0.855**	**0.0155**	0.49	0.41	−0.02	**16.02**
	IBP	0.819	0.0811	0.48	0.23	−0.04	6.56
	ProtoPNet	0.848	0.2964	0.37	0.1	−0.06	2.18
	ProtoTree	0.821	0.2122	0.43	0.17	**0.04**	13.75

for IC, where most values are very close to 0, meaning the saliency scores reflect the impact on the class score as much as uniformly random values. This highlights the fact that attention models and explanation methods are currently not designed for this objective, although it could be an interesting property. Globally, the metrics largely disagree on which method or model provides the most faithful explanations, which may be due to the chaotic behavior of the model when presented with the OOD inputs generated by the metrics.

4.5 Discussion

One common limitation of the metrics discussed in this chapter is that they are designed to evaluate single saliency maps, which makes their usage less straightforward for multi-part attention architectures, like B-CNN, BR-NPA, IBP, ProtoPNet, and ProtoTree. In our benchmark study, we selected the most important attention map, but to fully represent the model's behavior, all attention maps should be considered. Computing the mean attention maps would not be faithful to the model as each map has varying importance in the decision-making process. However, estimating the weight of each map for a pondered mean also raises challenges due to the diversity of architectures.

The low values of DC and IC in our benchmark do not imply unsatisfactory model and method performance but rather highlight that the calibration property has not been explored until now. To address this, we propose quantifying the sparsity and calibration of saliency maps, as these properties remain unstudied yet could be relevant for interpretability. However, determining their actual usefulness requires subjective testing in user experiments, which should involve various experimental setups based on the specific context, target audience, and purpose of the explanation.

The presentation of explanations to users is an open question, with different approaches used in literature, such as applying masks on the input image or superimposing the explanation over the image. Tasks given to users for evaluating explanations can also vary, ranging from predicting the network's prediction [21] to recognizing objects with masked input images [22]. By comparing objective metrics with user study rankings based on the impact of explanation methods on user understanding, we can determine which metrics best reflect the users' comprehension of the model.

However, current user studies suggest that providing saliency maps to users has little effect on their understanding and trust in the model. Though some studies show improvements in prediction with saliency maps, the effect size is small, raising questions about the significance of changing explanation methods in user perception. Further investigation and exploration of user studies are required to address these aspects and better understand the practical impact of explanation methods on users' interactions with interpretable models.

4.6 Related Works

In this section, we delve into prior research focusing on faithfulness metrics, particularly their design and assessment.

Faithfulness metric design: In this chapter, we focused our attention on multi-step metrics, i.e., metrics that apply several perturbations to the image and require several inferences. Other such metrics have been proposed in the literature, which have similarities with the metrics discussed here. For example, the "Faithfulness" metric [23] is similar to DC and IC in that it consists in measuring the correlation between classification score drops and saliency scores. The difference between DC and IC is that the "Faithfulness" metric does not apply cumulative perturbations to the image but rather only perturb one area at a time.

Another example is the Iterative Removal Of Features (IROF) metric that is also similar to DAUC [24]: After removing features in decreasing order, the metric is computed by measuring the area over the class score vs. saliency score curve. The difference with DAUC (and all the other metrics discussed here) is that it does not split the input image into a rectangular grid but rather computes super-pixels to obtain a list of regions. There also exist metrics that remove feature neither in an ascending nor descending order but rather perturb the image in a random fashion [25, 26].

A few single-step metrics have also been proposed. The Average Drop (AD) consists in applying the saliency map as a mask on the input image and measuring the average drop in confidence observed [3]. The Increase In Confidence (IIC) uses the same protocol but instead of measuring the confidence variation, it computes the number of images on which the masking increased the confidence of the model [3]. According to the authors, this masking procedure highlights the salient areas, and therefore the confidence of the model should increase. For this reason, the AD and IIC metrics should be respectively minimized and maximized to reflect an improvement. The last single-step metric we have identified is the Average Drop in Deletion (ADD), which reverses the saliency map before applying it to the image, resulting in the masking of the salient areas [2]. In this case, the classification confidence is expected to decrease, which is why this metric should be maximized.

Faithfulness metric evaluation: Previous work has already studied the properties of faithfulness metrics. Notably, Tomsett *et al.* investigated existing metrics and observed little consistency in how the metrics are computed in the literature and that these inconsistencies in implementation lead to significantly different conclusions [27]. Furthermore, they also conducted a series of statistical tests inspired by the psychometric research field and showed that the metrics are statistically unreliable and inconsistent. In this chapter, we provide one possible cause of this problem, which is the OOD nature of the samples passed to the model during the computation of the metrics. These samples could induce a chaotic behavior of the model, which would worsen the fidelity of the saliency maps.

4.7 Future Work

As future work, we propose several potential solutions to address the OOD issue raised in this chapter. First, it could be useful to change the patch replacement method. For example, one could replace a patch with a patch from another image belonging to a different class. Given that every patch would belong to the training manifold, it is likely that perturbed image would induce less severe OOD issues. Second, it would probably be helpful to train the model with perturbed images. By exposing the model to images similar to those generated during the computation of faithfulness metrics, the model could learn to effectively process perturbed inputs. Finally, further improvements may also be achieved by calibrating the model. A calibrated model, when presented with a substantially masked image, is more likely to recognize potential uncertainties and express lower confidence. Note that this strategy is effective only when the model has been trained on perturbed images.

4.8 Conclusion

In this chapter, we first studied two aspects of the DAUC and IAUC metrics. We showed using UMAP visualizations and an SVM classifier that they may generate OOD samples which might negatively impact their reliability. Also, we showed that they only take into account the ranking of the saliency scores and that the visual appearance of a saliency map can significantly change without the DAUC and IAUC metrics being affected. Then, we proposed to quantify two understudied aspects of saliency maps: the sparsity and the calibration (DC and IC). Furthermore, we provided context for our contribution by examining related works that assess the reliability of faithfulness metrics. Our discussion highlighted concerns regarding their inconsistent and unreliable nature that we primarily attributed to the OOD nature of samples processed during computation. Next, we discussed potential modifications to the model training and metrics design to address the OOD issue. Finally, we concluded with general remarks on the studied metrics, explored the issues of a user study to evaluate their usefulness, and reviewed related works on metric evaluation and design.

References

[1] V. Petsiuk, A. Das and K. Saenko, Rise: Randomized input sampling for explanation of black-box models (2018).

[2] H. Jung and Y. Oh, Lift-cam: Towards better explanations for class activation mapping (2021), *arXiv* abs/2102.05228.

[3] A. Chattopadhay, A. Sarkar, P. Howlader and V. N. Balasubramanian, Grad-cam++: Generalized gradient-based visual explanations for deep convolutional networks, in *2018 IEEE Winter Conference on Applications of Computer Vision (WACV)* (2018), pp. 839–847.

[4] S. Ghosh, R. Shet, P. Amon, A. Hutter and A. Kaup, Robustness of deep convolutional neural networks for image degradations, in *2018 IEEE International Conference on Acoustics, Speech and Signal Processing (ICASSP)*. IEEE (2018), pp. 2916–2920.

[5] C. Wah, S. Branson, P. Welinder, P. Perona and S. Belongie, The Caltech-UCSD Birds-200-2011 Dataset, California Institute of Technology, Technical Report CNS-TR-2011-001 (2011).

[6] K. He, X. Zhang, S. Ren and J. Sun, Deep residual learning for image recognition, in *2016 IEEE Conference on Computer Vision and Pattern Recognition (CVPR)* (2016), pp. 770–778.

[7] H. Wang, Z. Wang, M. Du, F. Yang, Z. Zhang, S. Ding, P. Mardziel and X. Hu, Score-cam: Score-weighted visual explanations for convolutional neural networks, in *Proceedings of the IEEE/CVF Conference on Computer Vision and Pattern Recognition Workshops* (2020), pp. 24–25.

[8] C. Guo, G. Pleiss, Y. Sun and K. Q. Weinberger, On calibration of modern neural networks, in D. Precup and Y. W. Teh, (eds.), *Proceedings of the 34th International Conference on Machine Learning*, 06–11 August 2017, Vol. 70. Proceedings of Machine Learning Research, pp. 1321–1330.

[9] J. Zhang, B. Kailkhura and T. Y.-J. Han, Mix-n-match: Ensemble and compositional methods for uncertainty calibration in deep learning, in H. Daumé III and A. Singh (eds.), *Proceedings of the 37th International Conference on Machine Learning*, 13–18 July 2020, Vol. 119. Proceedings of Machine Learning Research, pp. 11 117–11 128.

[10] J. Nixon, M. W. Dusenberry, L. Zhang, G. Jerfel and D. Tran, Measuring calibration in deep learning, in *CVPR Workshops*, Vol. 2, no. 7 (2019).

[11] R. R. Selvaraju, M. Cogswell, A. Das, R. Vedantam, D. Parikh and D. Batra, Grad-cam: Visual explanations from deep networks via gradient-based localization, in *2017 IEEE International Conference on Computer Vision (ICCV)* (2017), pp. 618–626.

[12] S. Desai and H. G. Ramaswamy, Ablation-cam: Visual explanations for deep convolutional network via gradient-free localization, in *2020 IEEE Winter Conference on Applications of Computer Vision (WACV)* (2020), pp. 972–980.

[13] T. Hu and H. Qi, See better before looking closer: Weakly supervised data augmentation network for fine-grained visual classification (2019), *CoRR* abs/1901.09891.

[14] T. Gomez, S. Ling, T. Fréour and H. Mouchère, Improve the interpretability of attention: A fast, accurate, and interpretable high-resolution attention model (2021).

[15] Z. Huang and Y. Li, Interpretable and accurate fine-grained recognition via region grouping (2020).

[16] C. Chen, O. Li, A. Barnett, J. Su and C. Rudin, This looks like that: Deep learning for interpretable image recognition, in *NeurIPS* (2019).

[17] M. Nauta, R. van Bree and C. Seifert, Neural prototype trees for interpretable fine-grained image recognition (2021).

[18] A. Paszke, S. Gross, S. Chintala, G. Chanan, E. Yang, Z. DeVito, Z. Lin, A. Desmaison, L. Antiga and A. Lerer, Automatic differentiation in pytorch, in *NIPS-W* (2017).

[19] O. Russakovsky, J. Deng, H. Su, J. Krause, S. Satheesh, S. Ma, Z. Huang, A. Karpathy, A. Khosla, M. Bernstein, A. C. Berg and L. Fei-Fei, ImageNet large scale visual recognition challenge, *International Journal of Computer Vision (IJCV)* **115**(3), pp. 211–252 (2015).

[20] T. Akiba, S. Sano, T. Yanase, T. Ohta and M. Koyama, Optuna: A next-generation hyperparameter optimization framework (2019).

[21] A. Alqaraawi, M. Schuessler, P. Weiß, E. Costanza and N. Berthouze, Evaluating saliency map explanations for convolutional neural networks: A user study. IUI'20. Association for Computing Machinery, New York (2020), pp. 275–285.

[22] D. Slack, A. Hilgard, S. Singh and H. Lakkaraju, Reliable *post hoc* explanations: Modeling uncertainty in explainability, *Advances in Neural Information Processing Systems*, Vol. 34 (2021).

[23] D. Alvarez-Melis and T. S. Jaakkola, Towards robust interpretability with self-explaining neural networks, in *Proceedings of the 32nd International Conference on Neural Information Processing Systems.* NIPS'18. Curran Associates Inc., Red Hook (2018), pp. 7786–7795.

[24] L. Rieger and L. Hansen, Irof: A low resource evaluation metric for explanation methods, in *Proceedings of the Workshop AI for Affordable Healthcare at ICLR 2020* (2020); Conference date: 26-04-2020 through 26-04-2020.

[25] U. Bhatt, A. Weller and J. M. F. Moura, Evaluating and aggregating feature-based model explanations, in *Proceedings of the 29th International Joint Conference on Artificial Intelligence, IJCAI'20* (2021).

[26] A.-P. Nguyen and M. R. Martínez, On quantitative aspects of model interpretability (2020), https://arxiv.org/abs/2007.07584.

[27] R. Tomsett, D. Harborne, S. Chakraborty, P. Gurram and A. Preece, Sanity checks for saliency metrics, *Proceedings of the AAAI Conference on Artificial Intelligence*, April 2020, Vol. 34, no. 4, pp. 6021–6029 (2020).

© 2025 World Scientific Publishing Company
https://doi.org/10.1142/9789811289125_0005

Chapter 5

Efficient Segmentation of E-Waste Devices With Deep Learning for Robotic Recycling

Cristof Rojas[*], Antonio Rodríguez-Sánchez[†], and Erwan Renaudo[‡]

Faculty of Mathematics, Computer Science and Physics, University of Innsbruck, 6020 Innsbruck, Austria
[*]*cristof.rojas@student.uibk.ac.at*
[†]*antonio.rodriguez-sanchez@uibk.ac.at*
[‡]*erwan.renaudo@uibk.ac.at*

Abstract

Recycling obsolete electronic devices (E-waste) is a dangerous task for human workers. Automated E-waste recycling is an area of great interest but challenging for current robotic applications. Robotics can help automate the E-waste recycling task and alleviate the hazards for human workers, but many challenges must be tackled. In this chapter, we focus on perception: how to segment E-waste devices into their composing parts so that the robot knows where to interact to disassemble the device and manipulate the parts. First, we extend a dataset of hard-drive disk (HDD) components with angled views of the devices. The new data contain labelled occluded and non-occluded points of view of the parts in order to represent the variety of poses the HDD can be in the actual task. Similarly, we complement the HDD dataset with a new dataset containing phone parts. By adding a new task, we represent the variety of devices the robot has to deal with. Both these additions contribute to increasing the quality of the learning data. We then perform an extensive evaluation with five different state-of-the-art models, namely CenterMask, Blend-Mask, SOLOv2, UNINEXT, and YOLOv8 (including variants), and two types of metrics: the average precision as well as the frame rate. We show that YOLOv8 outperforms the other models both in terms of precision and prediction rate. We also evaluate whether the progressive addition of the new phone task disturbs the performance of the model and show that although the intermediate accuracy on the new data is degraded, YOLOv8 is able to perform both tasks with the same performance as one. We conclude that instance segmentation using state-of-the-art deep learning methods can not only predict the complex shapes of a device's parts for robotic manipulation but also manage multiple devices at a high enough rate to give fast perceptive information to the robot.

5.1 Introduction

The ever-increasing volume of obsolete electronic devices (E-waste) is a strong incentive for the worldwide development of efficient recycling policies [1]. The existing recycling processes are mostly operated by humans, which causes health issues as well as a waste of rare materials [2,3]. Thus, the automation of the recycling process through the use of robots is a key step toward tackling these problems: Hazardous materials for humans are not harmful to robots, and the use of non-destructive methods allows the efficient collection of rare materials during extended working periods.

However, such a recycling process requires adaptive robotic manipulation on a highly diverse set of objects (e.g., domestic appliances, laptops, desktops, and professional-grade network equipment) and thus fast flowing, precise information on the object composition to disassemble them. The robot perception module must be able to identify the parts of the target object, its centre, orientation, and boundaries. This information lets the robot lever, grasp or push the parts away. Our previous work showed that this task can be formalized as an image segmentation problem and tackled by a convolutional neural network (CNN)-based deep neural network trained to predict the part class with a resolution to the pixel [4]. We showed that this problem can be tackled at a high frequency using a more compact, single-stage deep learning algorithm [5]. In addition, previous work focused on a hard-drive disk (HDD) dataset as a representative use case of the type of devices found in E-waste recycling applications. An evaluation has also been conducted with GPU parts and showed the system's generalization abilities, although only with a two-stage model [6].

We extend these results (a) by diversifying the task with a phone dataset in addition to the HDD to further assess the generalization ability of the system and (b) by evaluating two additional approaches than the work in [5], including UNINEXT [7], a unified model for diverse instance perception tasks, and YOLOv8 [8], a state-of-the-art real-time instance segmentation model, to widen our analysis on which approach is most promising to tackle such tasks.

The remainder of this chapter is organized as follows. Section 5.2 surveys existing approaches, for instance, segmentation and object tracking. Section 5.3 presents the extended HDD and phone dataset used in this study as well as the evaluated deep learning models. Section 5.4 reports the performance of the selected methods trained and tested on the dataset as well as the recorded inference time.

5.2 Related Work

Instance segmentation: Previous work [4] identifies various challenges that must be addressed when analyzing a disassembly scene. These include the need for accurate part detection to enable manipulation, the high degree of occlusion due to tight assembly, and the significant intra-class variability among device parts, which may be influenced by factors, such as brand, model, or potential damage. To solve this problem, the authors suggest formulating it as an instance segmentation problem and using deep learning methods, which are well suited for this task.

Deep learning methods utilize CNNs to achieve a pixel-level segmentation of images. One such method is mask R-CNN [9], which employs a region-based convolutional neural network (R-CNN) to predict segmentation masks for each instance. However, this method relies on a two-stage detector that employs a *detect-then-segment* approach, making it slow and unsuitable for applications that require real-time detections.

In the last years, there have been many studies of *single-stage* instance segmentation detection methods that have on-par precision to two-stage detectors with significantly lower inference time. Based on their average precision (AP) on the COCO *test-dev* dataset and their inference time, we chose these single-stage models for our use case: CenterMask [10], BlendMask [11], SOLOv2 [12], UNINEXT [7] and YOLOv8 [8].

Other notable single-stage detectors include the following:

- **YOLACT++** [13]: The instance segmentation process is divided into two simultaneous tasks: creating a collection of prototype masks and predicting per-instance mask coefficients. Additionally, the authors implemented a faster non-maximum suppression algorithm called Fast-NMS and incorporated deformable convolutions into the encoder's backbone network.
- **PolarMask++** [14]: The model predicts object outlines in polar coordinates, incorporating both instance segmentation (masks) and object detection (bounding boxes) within a unified framework. Additionally, the authors incorporated a refined feature pyramid into their approach.
- **SipMask** [15]: The authors presented a spatial preservation module, a pooling technique that retains the spatial data of an object by producing distinct sets of spatial coefficients for each sub-region enclosed within a bounding box.

Object Tracking: Object Tracking primarily concentrates on monitoring individual instances of cars and pedestrians, as seen in popular benchmarks like MOT [16]. The prevailing tracking strategy involves a *tracking-by-detection* approach, in which object instances from each frame are extracted and utilized for the tracking procedure. Two distinct categories of MOT algorithms exist: *batch* [17] and *online* methods [18]. In *batch tracking*, all frames, including future frames, are employed to identify the object instance at a given time (i.e., in a specific frame). Conversely, *online tracking* techniques can only leverage past and present information. The UNINEXT [7] unifies diverse instance perception tasks into a unified object detection model, including the instance-level tasks of instance segmentation and video-level multiple object tracking with segmentation tasks.

Instance segmentation for disassembly and recycling: Although instance segmentation is implemented in various domains, as noted in [4], to the best of our knowledge, very few works have explored the realm of automated E-Waste recycling. Outside the HDD use case in [4, 6], one of the rare exceptions is the work of

Sanderson [19], which employed deep learning networks to recycle flat panel displays. Additionally, the RecyBot project, introduced by [20], furnished a dataset of phone components with dense circuit boards that can be utilized for recycling purposes.

5.3 Methodology

5.3.1 *Instance segmentation*

Mask R-CNN [9], which is currently one of the state-of-the-art methods, for instance, segmentation, uses a *top-down* approach also called the *detect-then-segment* approach or the *two-stage* approach. It first detects the bounding boxes around the classes using a two-stage object detector and then creates the instance mask in each bounding box to differentiate the instances of the objects. Mask R-CNN is precise enough for the use case of recycling HDDs [4]. Therefore, its performance defines our baseline for precision, the drawback being that it has an inference time of only 20 FPS. To be considered a good replacement, the *single-stage* models must at least achieve an equivalent average precision with a higher inference speed. Our experiments (Section 5.4) show that the best model for our use case is one of the YOLOv8 [8] variants.

5.3.2 *Tracking*

We adopted a simple frame-by-frame approach for tracking, using only the information from the previous frame. This method proved to be sufficient for our purposes. In order to track the movement, we extracted the contours of each predicted instance mask and then calculated the midpoint of each contour. This approach provided a more accurate result than using the midpoint of the bounding boxes. The segmentation predictions produced bitmasks where 1 corresponded to the bits of the instance of one of the object's parts, and 0 corresponded to the rest of the image. We determined the outermost contours of each bitmask separately and calculated the midpoint using a weighted average of the pixels for each contour. In cases where multiple instances of the same part were detected, we chose the part with the smallest euclidean distance between the midpoints.

We used a sequence of images of a hard-drive during movement to assess the tracking performance of the selected models on our dataset. We calculated the average precision (AP) for each image in the sequence to evaluate the accuracy of the HDD tracking during movement.

Other possibilities include using the UNINEXT [7] model, which is a universal model that can be used for multiple tasks, including the task of multi-object tracking and segmentation delivering state-of-the-art results on the BDD100k dataset [21]. The YOLOv8 [8] model integrates different tracking algorithms, such as the BoT-SORT tracker [22].

5.3.2.1 BlendMask

BlendMask is an instance segmentation method that adopts a *detect-then-segment* approach, similar to Mask R-CNN, where an object detector generates instance proposals and then a sub-network predicts the instances. It simplifies the head of Mask R-CNN by reusing a predicted global segmentation mask, which is achieved by using a tensor-product operation called *Blend*. BlendMask is built upon FCOS [23], a fully convolutional single-stage object detection framework. For instance segmentation, BlendMask adds lightweight top and bottom modules adopted from the top-down and bottom-up approaches. Top-down approaches predict the entire instance based on features, while bottom-up approaches group pixels into instances to create local predictions [24]. BlendMask merges these approaches, where the bottom module predicts a set of object bases from the backbone and the top module predicts attention maps, which represent the object's coarse shape and pose. Finally, the blender module combines the position-sensitive bases according to the attention to produce the final segmentation.

5.3.2.2 SOLOv2

SOLO [25] stands for segment objects by locations and is an instance segmentation framework that divides an input image into uniform grids. If the centre of an object is inside the current grid cell, the semantic class is predicted, and the object's instance is segmented. Segmenting the object's instance and predicting its semantic class are performed simultaneously. SOLO also introduces "instance categories" to assign categories to each pixel within an instance based on the instance's location and size. Unlike the *detect-then-segment* approach, SOLO is a single-stage instance segmenter that uses a direct approach, benefiting from the advantages of fully connected networks, and outputs instance masks and their class probabilities directly. SOLOv2 [12] improves upon SOLO by using a dynamic scheme, where the mask generation process is divided into a mask kernel prediction and mask feature learning, and appropriate location categories are assigned to different pixels. SOLOv2 also reduces inference time by introducing a matrix non-maximum suppression (NMS) technique with parallel matrix operations.

5.3.2.3 UNINEXT

Universal instance perception as object discovery and retrieval (UNINEXT) [7] is a universal instance perception model that reformulates different instance perception tasks into a unified object discovery and retrieval paradigm. It can perceive different types of objects by changing the input prompts and benefits from enormous data from different tasks and label vocabularies for jointly training general instance-level representations. The unified model is parameter-efficient and can handle multiple tasks simultaneously, including image-level tasks, such as instance segmentation and video-level object tracking tasks like multiple object tracking. Additionally, it has vision-and-language tasks, such as referring expression segmentation.

UNINEXT uses an encoder–decoder architecture and deformable DETR's multi-scale deformable self-attention to exchange target information from different scales. The input of the decoder consists of multi-scale features, object queries, and reference points. The prediction heads produce both boxes and masks of the targets. The proposed retrieval head selects objects by prompt-instance matching, enabling UNINEXT to learn universal instance representations.

5.3.2.4 YOLOv8

You only look once (YOLO) [26] has been a popular framework for many applications in regard to real-time object detections, with an extraordinary balance of speed and accuracy. In the last years, many iterations have evolved, each building upon the previous implementation by enhancing its performance, where the main focus has been on real-time object detection. For our case, we chose the latest iteration YOLOv8 [8] with the so far best results in AP and inference speed. YOLOv8 is one of the best real-time instance segmentation models in terms of the trade-off between AP and inference time on the COCO dataset [27].

YOLOv8 is anchor-free, which means that it reduces the number of box predictions and has sped up the non-maximum suppression (NMS) technique. Similar to its predecessors, YOLOv8's architecture consists of a backbone, head, and neck where the convolutional layers in the backbone have been improved. The backbone is responsible for extracting useful features from the input image using CNNs. The neck connects the backbone to the head and improves the representation of the features, e.g., by using additional convolutional layers and other mechanisms. Finally, the head is responsible for making the predictions based on these features and performing the instance segmentation using subnetworks.

YOLOv8 is implemented with different backbones: nano, small, medium, large, and extra-large scalings. The nano version has a relatively high drop of 4% accuracy with a gain only of under 0.26 ms [8]. Furthermore, there is no significant difference between the large and extra-large scales. Therefore, for our experiments, we used the small (S), medium (M), and extra-large (X) scalings. In the end, YOLOv8 uses mosaic augmentation during training, which involves stitching together four images, forcing the model to learn the objects in new locations with partial occlusion against different surrounding pixels. However, it is turned off for the last 10000 iterations of training, as it has been shown to reduce performance if used for the complete training iteration.

5.3.3 Training

The models were trained for 20000–30000 iteration, and all the models were pre-trained on the COCO dataset. The precision and inference time of the models depend highly on the resolution of the input image: a higher resolution image leads to results of higher precision at the cost of a higher inference time and vice versa. We train some models in higher and lower resolution input images to evaluate the resulting trade-off between precision and inference time. To increase the speed performance of BlendMask, SOLOv2, and CenterMask, we trained one version with lower resolution input images. YOLOv8 was trained on a single resolution with different backbone scalings, while UNINEXT was trained on a single resolution and backbone. Furthermore, all the models have at least the augmentations of random lightning, random brightness, and random horizontal flips.

5.3.4 *Precision*

The precision of the models is evaluated using the COCO [27] evaluation metric. This method, known as average precision (AP), measures the accuracy for object detection and instance segmentation by calculating the intersection over union (IoU) defined in Eq. (5.1):

$$\text{IoU} = \frac{\text{area}(\text{Mask}_p \cap \text{Mask}_{gt})}{\text{area}(\text{Mask}_p \cup \text{Mask}_{gt})}. \tag{5.1}$$

Mask_p and Mask_{gt} correspond to the predicted mask and ground-truth mask, respectively, and the IoU (Eq. 5.1) calculates the intersection between the ground-truth mask and the predicted mask. The predicted mask is correctly detected if the IoU is equal to or greater than a given threshold (e.g., 50% for $\text{AP}_{\text{IoU}=.5}$). AP is the area under the precision and recall curve of 10 IoU values from 50% to 95% in 5% increments.

5.3.5 *Inference time*

We define the *inference time* of a network as the duration between presenting an input image and obtaining the prediction output. The average framerate that the network can achieve is then deduced by calculating this time for the entire test dataset. It is noteworthy that this metric is highly reliant on the hardware employed. While Jetson computers, such as the Jetson Xavier, are becoming increasingly popular in embedded robotic systems, we utilized a desktop equipped with an RTX 3090 GPU for our stationary recycling system. As a result, all timing measurements in this study were obtained using a single RTX 3090 GPU with a batch size of 1.

Additionally, we evaluated the number of parameters and floating-point operations (GFlops) for the models with the best results.

5.3.6 *Dataset*

We use a dataset of annotated HDD and phone parts separated into 21 classes. Each of these part types has to be recognized by the system in order to be recycled differently. The original dataset had 500 HDD images, containing only top views of the hard drives. This limits the system, as it cannot recognize the parts with good accuracy when the hard drive is rotated or moved by the robot.

Our extended dataset contains an additional 400 annotated HDD images, including tilted views of the device and 524 phone images provided by [20]. Figure 5.1 shows the examples of the phones and hard drives in top and angled views, and the number of instances for each part is listed in Table 5.1. The angled views were created with the HDDs rotated around the central axis along their length in a variety of angles and disassembly stages (Figures 5.1(c) and 5.1(d)). For the purpose of this study, we believe that the current tilting angles suffice, even though additional angles could be incorporated. The present angles encompassed most of the scenarios encountered by the robot. The ground-truth annotations in the dataset were manually

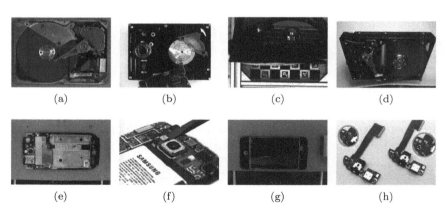

Fig. 5.1. Example of different HDD and phone images [20]: (a) HDD top view; (b) HDD top-view disassembly stage; (c) HDD angled-view train dataset; (d) HDD angled-view test dataset; (e) phone opened; (f) phone opened angled; (g) phone screen; (h) phone charging port.

Table 5.1. Number of images and instances of our train and test dataset.

Taxonomy name	Train instances	Test instances	Total
Magnets	265	70	335
Flexible Printed Circuit	199	93	292
Read/Write-Head	246	61	307
Spindle Hub	301	101	402
Platters Clamp	220	70	290
Platter(s)	267	124	391
Drive Bay	612	228	840
The Drivers Lid	197	78	275
PCB	154	51	205
Head Contacts	42	45	87
Top Dumper	11	40	51
Battery	110	11	121
Charging Port	53	10	63
Motor	89	1	90
Cover	71	13	84
Back Cover	19	1	20
Screen	17	1	18
Cable	314	40	354
Camera	131	20	151
Connector	312	44	356
Total instances	**3754** (77%)	**1120** (23%)	**4874**
Total images	**1165** (80%)	**298** (20%)	**1463**

Note: The name of taxonomy refers to the part names. The horizontal line separates the HDD from the phone parts.

labeled using the online tool makesense.ai.[a] Refer to the IMAGINE project website[b] for a comprehensive taxonomy of HDD parts.

The HDD dataset comprises images of seven distinct brands and various models featuring damaged devices and different stages of disassembly. The most frequently appearing component in the images is the drive bay, visible in nearly all images, while the top dumper and head contacts are the least occurring parts. The top dumper is specifically designed to safeguard the platter and is only incorporated in a limited number of HDDs. Head contacts serve as the link between the

---

[a] https://www.makesense.ai/.
[b] https://imagine-h2020.eu/hdd-taxonomy.php.

PCB and R/W-Head, and are often concealed behind the PCB on the backside, becoming visible only after the PCB is removed. The rest of the components are represented in roughly equal quantities.

Additionally, we added the phone dataset described in [20], which was originally designed for semantic segmentation. This dataset includes phones from 10 different disassembled phone brands. However, we had to distinguish between different instances of each category, so we modified the dataset to include unique instance IDs for each object. This involved separating masks that were assigned to multiple instances of the same class, which only affected some of the cover and motherboard categories. We removed the screw category dataset to streamline our dataset further since our system already includes a specialized screw detector. We also removed images from the training set with transformations such as rotations, random noise, Gaussian blur, sharpness, and changes in lightness, as we had our own augmentations of random lightning, brightness, horizontal flips, and others, depending on the model to train. Finally, we added instances that were missing from the dataset to improve the overall completeness and accuracy of the data. By making these adjustments, we were able to create a more effective and efficient dataset that improved the performance of our instance segmentation model.

Despite the relatively low number of images in the dataset for deep learning, we achieve reasonably good results (over 60% average precision, Section 5.4), which is due to the lower variance between certain parts between different brands and models, such as the platter, spindle hub, platters clamp, drive bay, drivers lid, camera, battery, and connectors. We noticed that certain parts would benefit from more instances in the dataset. For example, due to its higher variability and low occurrence, the top dumper has significantly lower average precision than the other parts at only around 50%. The evaluated neural networks have been pre-trained on the COCO dataset [27].

5.4 Results

5.4.1 *Precision and speed*

Mask R-CNN is chosen as the baseline precision that the model has to reach, as it was shown to be precise enough to allow disassembling

HDDs [4]. Mask R-CNN combines a residual neural network [28] (R101) with a feature pyramid network (FPN) [29] backbone and achieves an AP of 65.5% with an inference speed of about 20 FPS. These results are set as the minimum requirements for the remaining models. The state-of-the-art models evaluated on our dataset are CenterMask-Lite [10], BlendMask [11], SOLOv2 [12], UNINEXT [7] and YOLOv8 [8].

Figure 5.2 shows the trade-off between precision and speed: deeper backbones consistently result in higher precision, compared to the shallower ones with lower precision but higher inference speed. Additionally, models with lower input image sizes had lower AP but higher inference speed.

UNINEXT achieves the highest AP with the R50 backbone. However, it has a lower inference speed than the Mask R-CNN model. YOLOv8 is the fastest model by a large margin with over 200 FPS,

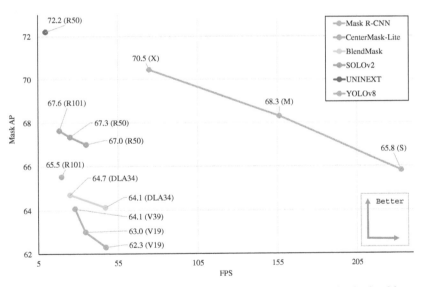

Fig. 5.2. Trade-off between Speed and AP for the various methods, backbones, and input image sizes. The labels of each marker indicate the resulting AP and used backbone are in parentheses. The evaluated shorter side of the input image was resized to 512 for BlendMask DLA34, 512 and 800 for SOLOv2, 600 and 800 for CenterMask V19-FPN, 800 for the V39-FPN backbone, 800 for UNINEXT and 640 × 640 for YOLOv8. The models with smaller image sizes and shallower backbones always result in higher inference speed but lower AP.

10 times the speed of Mask R-CNN, on the smaller backbone (S). SOLOv2 with the R50 and R101 backbones and BlendMask with the deep layer aggregation (DLA34) [30] backbone is generally faster than Mask R-CNN. However, BlendMask has lower precision compared to Mask R-CNN. CenterMask-Lite with VoVNet [31] (V19-FPN and V39-FPN) backbone has lower precision than BlendMask and about the same inference speed.

For a more complete evaluation, we added the deeper backbones R101 and V39 to see which options provide the best precision and speed trade-off. Our results indicate that for our use case, the shallower backbones are better suited, as the improvement in speed is more significant than the improvement in precision. The best option is the YOLOv8 model: it achieves the highest AP of all the models faster than Mask R-CNN. The extra-large backbone has more than 2% AP compared to SOLOv2 and up to 5% compared to Mask R-CNN while being significantly faster than all of these models.

We trained the YOLOv8 model with the extra-large (X) backbone on an incrementally increasing number of images from the phone dataset to evaluate the improvement gained by adding more data to the system. This experiment allows the evaluation of the amount of retraining needed to add a new task (phone parts classification) when the network has been trained on the HDD dataset. For each iteration, the model was retrained by randomly adding 0%, 25%, 50%, 75%, and 100% of images from the phone dataset. The results are shown in Figure 5.3. Using no data from the phone dataset, the model naturally did not correctly detect any of its parts. For each increase in the dataset's size, the model's AP increased. The few exceptions correspond when the random selection of images excluded certain classes, e.g., at the 50% data increase step: no screen appeared in the train dataset and therefore could not be detected. As expected, the most difficult classes to predict are the FPC and top dumper in the HDD dataset and the charging port, cables, and connectors in the phone dataset.

The results from [6] showed that the progressive addition of GPU to a model had generalization abilities. This was the case because HDDs and GPUs have overlapping classes, although they may still look different. However, in our case, the HDD and phone datasets don't have any overlapping classes and, therefore, had to at least partially be trained to detect any of the parts.

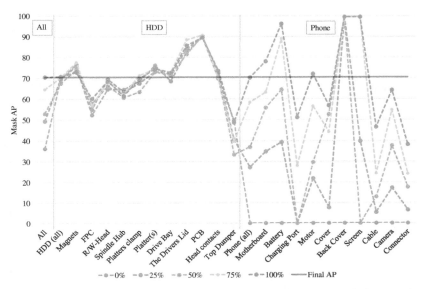

Fig. 5.3. Comparison of YOLOv8 with an extra-large backbone (X) trained on randomly chosen images from the phone dataset with increasing amounts. It was trained using 0%, 25%, 50%, 75%, and 100% of the phone dataset and evaluated every time using the complete test dataset. "All" is the AP of all parts, "HDD (all)" from all HDD parts, and "Phone (all)" from all the phone parts. The label "Final AP" is the AP of the model after the final training iteration using the complete data.

As expected, using more training significantly increases the performance of the phone dataset. Furthermore, the results show that the performance on the HDD dataset also slightly improved with the increase of the phone dataset, which can be seen in the "HDD (all)" column. The results also demonstrate that the overall final AP was approximately the same for the HDD parts "HDD (all)" and the phone parts "Phone (all)", indicating that this task on a different dataset has the same difficulty level. Overall, the model can progressively integrate a new task (thus, a new device with parts to segment) without losing precision.

5.4.2 Tracking

Following, we evaluate the precision of the faster models on images of HDDs in motion. Figure 5.4 displays an example sequence of frames,

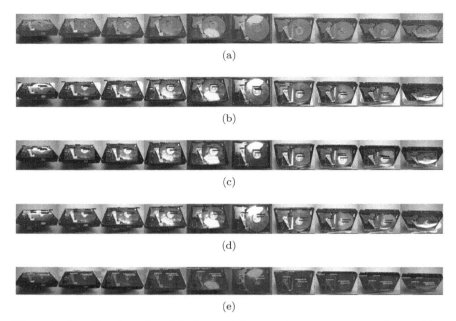

Fig. 5.4. Predicted masks with the models using lower image sizes for tracking evaluation. Sample images part of our test dataset, the top row corresponds to the ground-truth masks, and the models are the same as in Figure 5.2. Next to each predicted part, the class name and the confidence score is shown. We can already see the difference in AP between each model, where SOLOv2 R50-FPN has the most accurate masks for these images. (a) Ground truth; (b) CenterMask-Lite V19-FPN; (c) BlendMask DLA34; (d) SOLOv2 R50-FPN; (e) YOLOv8 X.

depicting different models' ground-truth and predicted masks. These images were taken on a simple mount that could change its angle (Figure 5.1(d)) and cast a shadow on the HDD, which was erroneously detected as part of the drive bay by CenterMask-Lite, SOLOv2, and YOLOv8, while BlendMask was more robust in detecting the drive bay, although it tended to detect it as smaller than it actually is. The test dataset used for evaluation had different conditions from the ones used in training to assess the models' robustness. The training images were taken with the camera mounted on the disassembly robot, while the test images were taken on a simple mount. BlendMask demonstrated greater robustness in detecting the drive bay, although it generally detected a smaller drive bay than it actually is.

Efficient Segmentation of E-Waste Devices 139

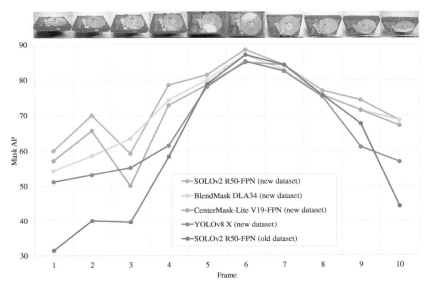

Fig. 5.5. AP on the frames of Figure 5.4. The images over the diagram are the corresponding ground-truth masks. In the images where the HDD is angled more, the AP of all the models decrease. SOLOv2 has the highest AP for all the frames with over 60%. Furthermore, we compare our extended dataset to the old dataset [4] by additionally training the SOLOv2 R50-FPN model solely on the older dataset that only contained top views. We can see that the SOLOv2 model trained on the old dataset has significantly worse AP for the images where the HDD is angled more, while the AP for the top-view images has about the same AP.

The precision of the faster models on images of HDDs during motion is analyzed in the following. Figure 5.5 illustrates the corresponding AP for each frame in Figure 5.4. The figure indicates that the models have lower AP for more angled HDDs, but the precision remains above 50% for all frames and models, with SOLOv2 yielding about 60% as the lowest AP result. Although YOLOv8 generally has the best results over the complete test dataset, it struggles with angled views of the HDD dataset. This indicates that YOLOv8 would highly benefit from using more images of HDDs in different views and angles.

The decrease in performance with higher angles is possibly due to the imbalance between top and angled views in the dataset, which has a majority of parts in top view (Figure 5.1(b)). This is because multiple disassembly stages with different inner component configurations are visible in top views, whereas these configurations are not visible in the most angled views, resulting in a natural imbalance in the dataset toward top views. By comparing the AP results of a SOLOv2 R50-FPN model trained on an older dataset with only top views of HDDs, we can observe that the old model performs significantly worse on higher angled views, particularly on the first and last frames where the HDD had wider angles. A small drop in precision is also visible in frame 3 compared to nearby frames due to the FPC part not being detected in this frame by SOLOv2, BlendMask or YOLOv8. The FPC is generally difficult to detect when the HDD is angled since its appearance changes more from different perspectives compared to the other parts in this dataset. Thus, expanding the dataset with more variations of FPCs in different orientations would be advantageous.

Table 5.2 provides a detailed analysis of the models, focusing on the number of parameters and GFlops. The number of parameters (in millions) remains constant regardless of the input image size. YOLOv8 has the largest number of parameters, while SOLOv2 with an R101-FPN backbone has a lower total number of parameters than Mask R-CNN with the same backbone. In contrast, BlendMask has the fewest parameters, while YOLOv8 has the most. GFlops represent the number of operations required to run a single instance of the given model, and for our study, we averaged the GFlops over 100 input images.

Table 5.2. Number of parameters and GFlops per model.

Backbone	Mask R-CNN R101	BlendMask DLA-34	SOLOv2 R101	SOLOv2 R50	YOLOv8 X
Image size	800 × *	512 × *	512 × *	512 × *	640 × 640
Total par. [M]	63.0	25.3	65.2	46.5	71.7
Backbone par. [M]	45.7	17.3	45.7	26.8	30.9
GFlops	250.4	86.1	289.3	132.8	344.6
FPS	20	47	19	35	89

Note: The number of parameters in millions and the number of GFlops. YOLOv8 has the most number of parameters and GFlops, while BlendMask has the least.

5.5 Conclusion

In this chapter, we extended the dataset for the detection of device parts in an E-waste recycling scenario by adding HDD images with different angles and a phone part dataset. With this dataset, we compared several state-of-the-art single-stage instance segmentation models and compared them to Mask R-CNN. The best results were achieved by using the single-stage YOLOv8 model, which achieved 70.5% AP while achieving an inference speed of 75 FPS on our test dataset. These results have higher precision than Mask R-CNN (used in the preliminary work) and three times the speed. Lower scaling backbone achieved on-par AP with Mask R-CNN while achieving over 11 times speed increase.

The predicted masks were employed to track the movement of an HDD during disassembly, and the tracking precision was evaluated by calculating the AP of the HDD during motion. The results indicated that our approach achieved an AP of at least 60% even on images with a high level of occlusion on different parts. Additionally, we trained the YOLOv8 model with the extra-large (X) backbone on an incrementally increasing number of images. The results showed that the model can progressively integrate a new task (thus a new device with parts to segment) without losing precision. To further assess the effectiveness of our extended dataset, we trained the SOLOv2 model on the old dataset and compared the AP of the HDD during motion. The comparison demonstrated that our extended dataset significantly improved the precision of images with wider angles and higher levels of occlusion. These findings suggest that single-stage deep learning-based instance segmentation is a promising approach for an online setup without compromising precision. Therefore, this approach holds great potential for developing an autonomous and adaptive recycling system for E-waste processing.

Acknowledgements

We acknowledge the European Union's Horizon 2020 program for grant agreement no. 731761 (IMAGINE) as well as the French Agence Nationale de la Recherche (ANR) and the Austrian Science Fund (FWF) for the joint grant (ANR-21-CE33-0019-01/FWF: I 5755-N, ELSA).

References

[1] K. Liu, Q. Tan, J. Yu and M. Wang, A global perspective on e-waste recycling, *Circular Economy* **2**, 1, p. 100028 (2023), doi:10.1016/j.cec.2023.100028, https://www.sciencedirect.com/science/article/pii/S2773167723000055.

[2] J. Yang, J. Bertram, T. Schettgen, P. Heitland, D. Fischer, F. Seidu, M. Felten, T. Kraus, J. N. Fobil and A. Kaifie, Arsenic burden in e-waste recycling workers – A cross-sectional study at the Agbogbloshie e-waste recycling site, Ghana, *Chemosphere* **261**, doi:10.1016/j.chemosphere.2020.127712.

[3] P. United Nations Environment Programme, I. ITU and U. N. U. UNIDO, A new circular vision for electronics time for a global reboot (2019).

[4] E. Yildiz, T. Brinker, E. Renaudo, J. Hollenstein, S. Haller-Seeber, J. Piater, and F. Wörgötter, A visual intelligence scheme for hard drive disassembly in automated recycling routines, in *Proceedings of the International Conference on Robotics, Computer Vision and Intelligent Systems — ROBOVIS*, INSTICC. SciTePress, pp. 17–27 (2020).

[5] C. Rojas, A. Rodríguez-Sánchez and E. Renaudo, Deep learning for fast segmentation of e-waste devices' inner parts in a recycling scenario, in *International Conference on Pattern Recognition and Artificial Intelligence*. Springer, pp. 161–172 (2022).

[6] E. Yildiz, E. Renaudo, J. Hollenstein, J. Piater and F. Wörgötter, An extended visual intelligence scheme for disassembly in automated recycling routines, in *International Conference on Robotics, Computer Vision and Intelligent Systems, International Conference on Robotics, Computer Vision and Intelligent Systems*. Springer, pp. 25–50 (2022).

[7] B. Yan, Y. Jiang, J. Wu, D. Wang, P. Luo, Z. Yuan and H. Lu, Universal instance perception as object discovery and retrieval (2023), arXiv:2303.06674 [cs.CV].

[8] G. Jocher, A. Chaurasia and J. Qiu, YOLO by Ultralytics (2023), https://github.com/ultralytics/ultralytics.

[9] K. He, G. Gkioxari, P. Dollár and R. B. Girshick, Mask R-CNN, *CoRR* (2017), http://arxiv.org/abs/1703.06870.

[10] Y. Lee and J. Park, Centermask: Real-time anchor-free instance segmentation, *CoRR* (2019), http://arxiv.org/abs/1911.06667.

[11] H. Chen, K. Sun, Z. Tian, C. Shen, Y. Huang and Y. Yan, Blendmask: Top-down meets bottom-up for instance segmentation, *CoRR* (2020), http://arxiv.org/abs/2001.00309.

[12] X. Wang, R. Zhang, T. Kong, L. Li and C. Shen, Solov2: Dynamic, faster and stronger, *CoRR* (2020), https://arxiv.org/abs/2003.10152.
[13] D. Bolya, C. Zhou, F. Xiao and Y. J. Lee, Yolact++: Better real-time instance segmentation, *IEEE Transactions on Pattern Analysis and Machine Intelligence*, p. 1–1 (2020).
[14] E. Xie, W. Wang, M. Ding, R. Zhang and P. Luo, Polarmask++: Enhanced polar representation for single-shot instance segmentation and beyond, *CoRR* (2021), https://arxiv.org/abs/2105.02184.
[15] J. Cao, R. M. Anwer, H. Cholakkal, F. S. Khan, Y. Pang and L. Shao, Sipmask: Spatial information preservation for fast image and video instance segmentation, *CoRR* (2020), https://arxiv.org/abs/2007.14772.
[16] L. Leal-Taixé, Multiple object tracking with context awareness, *CoRR* (2014), http://arxiv.org/abs/1411.7935.
[17] J. Son, M. Baek, M. Cho and B. Han, Multi-object tracking with quadruplet convolutional neural networks, in *2017 IEEE Conference on Computer Vision and Pattern Recognition (CVPR)*, pp. 3786–3795 (2017).
[18] J. Xiang, G. Zhang and J. Hou, Online multi-object tracking based on feature representation and bayesian filtering within a deep learning architecture, *IEEE Access* **7**, pp. 27923–27935 (2019).
[19] A. Sanderson, *Intelligent Robotic Recycling of Flat Panel Displays*, Master's thesis, University of Waterloo (2019), http://hdl.handle.net/10012/14730.
[20] A. Jahanian, Q. H. Le, K. Youcef-Toumi and D. Tsetserukou, See the e-waste! training visual intelligence to see dense circuit boards for recycling, in *Proceedings of the IEEE/CVF Conference on Computer Vision and Pattern Recognition (CVPR) Workshops* (2019).
[21] F. Yu, H. Chen, X. Wang, W. Xian, Y. Chen, F. Liu, V. Madhavan and T. Darrell, Bdd100k: A diverse driving dataset for heterogeneous multitask learning, (2020), arXiv:1805.04687 [cs.CV].
[22] N. Aharon, R. Orfaig and B.-Z. Bobrovsky, Bot-sort: Robust associations multi-pedestrian tracking (2022), arXiv:2206.14651 [cs.CV].
[23] Z. Tian, C. Shen, H. Chen and T. He, FCOS: fully convolutional one-stage object detection, *CoRR* (2019), http://arxiv.org/abs/1904.01355.
[24] D. Neven, B. D. Brabandere, M. Proesmans and L. V. Gool, Instance segmentation by jointly optimizing spatial embeddings and clustering bandwidth, *CoRR* (2019), http://arxiv.org/abs/1906.11109.
[25] X. Wang, T. Kong, C. Shen, Y. Jiang and L. Li, SOLO: Segmenting objects by locations, *CoRR* (2019), http://arxiv.org/abs/1912.04488.

[26] J. Redmon, S. Divvala, R. Girshick and A. Farhadi, You only look once: Unified, real-time object detection (2016), arXiv:1506.02640 [cs.CV].

[27] T.-Y. Lin, M. Maire, S. Belongie, J. Hays, P. Perona, D. Ramanan, P. Dollár and C. L. Zitnick, Microsoft coco: Common objects in context, in D. Fleet, T. Pajdla, B. Schiele and T. Tuytelaars (eds.), *Computer Vision – ECCV 2014*. Springer International Publishing, Cham, pp. 740–755 (2014).

[28] K. He, X. Zhang, S. Ren and J. Sun, Deep residual learning for image recognition, *CoRR* (2015), http://arxiv.org/abs/1512.03385.

[29] T. Lin, P. Dollár, R. B. Girshick, K. He, B. Hariharan and S. J. Belongie, Feature pyramid networks for object detection, *CoRR* (2016), http://arxiv.org/abs/1612.03144.

[30] F. Yu, D. Wang and T. Darrell, Deep layer aggregation, *CoRR* (2017), http://arxiv.org/abs/1707.06484.

[31] Y. Lee, J. Hwang, S. Lee, Y. Bae and J. Park, An energy and gpu-computation efficient backbone network for real-time object detection, *CoRR* (2019), http://arxiv.org/abs/1904.09730.

Chapter 6

Shop Signboard Detection Using the ShoS Dataset

Mrouj Almuhajri[*,‡] and Ching Y. Suen[†,§]

[*]Saudi Electronic University, Saudi Arabia
[†]Concordia University, Canada
[‡]m.almuhajri@seu.edu.sa
[§]suen@cse.concordia.ca

Abstract

This chapter was motivated by the potential to facilitate impactful applications for both individuals and municipal agencies through the utilization of deep learning techniques for shop signboard detection. In the era of data-driven technologies, data play an important role to formulate the foundation of successful AI-based applications. Nevertheless, providing comprehensive datasets with meticulous annotation remains a challenge, requiring precise refinement and characterization to ensure optimal utilization. The primary objective of this chapter is to describe the process of collecting the Shop Signboard (ShoS) dataset from Google Street Views for the purpose of detecting signboards of shops on streets. A full set of 10k storefront signboards was captured and fully annotated comprising 7,500 images. Subsequently, the ShoS datasets underwent rigorous testing by running different baseline methodologies employing one-stage detectors YOLO and SSD. Among the various methods and configurations tested, YOLOv3 emerged as the top-performing model, achieving remarkable performance of 94.23% (mAP@0.5). The obtained results have been discussed, and potential avenues for further research and development have been proposed.

6.1 Introduction

Despite the vast availability of street view images provided by Google, the process of collecting storefront images and annotating them is challenging and time-consuming. This task is susceptible to human errors due to its intensive nature. The literature highlights that the detection and classification of storefronts from Google view images are frequently hindered by the absence of annotations or the presence of inaccurate ones [1–3]. The motivation behind this chapter stems from the potential to facilitate beneficial applications for municipal agencies, individuals, and other transportation-related applications.

Municipal government agencies enact policies and regulations to oversee the design of storefronts and their corresponding signboards. While human inspectors traditionally verify compliance with these policies and guidelines, AI-based systems offer the potential to expedite and enhance the accuracy of tasks related to the identification, classification, and assessment of compliance with shop signboard regulations. As an illustrative example, the city of Westmount in Quebec has implemented regulations governing the design aspects of storefronts and their signboards. These regulations encompass considerations, such as language, size, lettering, and graphic elements [4].

Detecting shop signboards is the foundational step for unlocking various potential applications including store classification. This capability holds the promise of enhancing the exploration of neighborhoods and identifying points of interest for diverse user groups, such as tourists and individuals with visual impairments, who can leverage smart devices for enhanced navigation and information retrieval.

The primary contributions of this chapter can be summarized as follows: (1) provide the ShoS dataset for the public for several research purposes including but not limited to shop signboard detection and (2) present outcomes derived from diverse baseline methodologies and applications applied to the ShoS dataset. This includes the utilization of YOLO and SSD models, offering insights into the effectiveness of these methodologies for shop signboard detection.

6.2 Related Work

6.2.1 *Street view imagery object detection*

In [2], the authors aimed to detect the whole storefronts given street panoramic views using the MultiBox model [5] which uses a single CNN based on GoogLeNet [6] with a 7×7 grid. A coarse sliding window mechanism was applied followed by non-maximum suppression to make fewer predictions of bounding boxes. This chapter found difficulty annotating that huge number of the dataset, so they used less amount of data in testing. For evaluation, they compared their work with selective search, multi-context heat MCH map, and human performance. Results revealed that MultiBox surpasses the first two methods as MultiBox gets a recall of 91% compared to 62% for selective search. Moreover, MCH could not detect the boundary of storefronts precisely. This was due to the fact that storefronts are more exposed to noise and they can abut each other.

Similarly, another recent study [3] proposed a system composed of several models for detecting the whole storefront from street views and classify it in further models. The detector was based on YOLOv3 [7] as they removed the layers responsible for detecting small objects. The authors evaluated their work using their limited dataset and compared it to SSD [8] and Faster RCNN [9]. Their methodology got mAP@0.50 = 79.37% which outperforms the others with 74.3% and 78.9%, respectively.

For that, some research results have been published in a similar area which is License Plate Detection (LPD). A lot of attention has been devoted to LPD with the enhancement of real-time object detection models. In [10], the first two models of YOLO [11] and [12] were used in a multi-stage manner to detect vehicles first using Fast-YOLO with IoU set to 0.125 and then the detected vehicle is cropped and fed into YOLOv2 network with an extremely low threshold to make sure they got all possibilities within the vehicle patch. Thus, the box with a higher confidence score is considered. This way, they were able to get high precession and recall nearly to 99%. In addition, in [13], the authors introduced a YOLO-inspired solution for LPD system as they trained and tested several models with different hyper parameter values in order to save computational processing time and power. A recall ratio of 98.38% is achieved outperforming two commercial products (OpenALPR and Plate Recognizer) significantly.

Another work [14] focused on providing a small and fast model for LPD so that it could work on embedded systems. The proposed system is based on Mobilenet-SSD (MSSD) to detect license plates. The authors further optimized the system by introducing feature fusion methods on the MSSD model in order to extract context information and hence better detection. The results revealed that their proposed system is 2.11% higher than the MSSD in terms of precision, and it is also faster than MSSD by 70 ms.

The progress in real-time object detection algorithms has witnessed significant acceleration, exemplified by the iterative releases of the YOLO model, which has undergone five versions as of this chapter [7, 11, 12, 15, 16]. Each version retains its utility, contributing distinct features that may, but not necessarily, enhance overall performance. In this research, YOLOv3 was adopted due to its commendable performance.

While substantial strides have been made in the field of License Plate Detection (LPD), the dynamics differ when applied to shop signboards due to their diverse shapes and forms in contrast to license plates. This visual heterogeneity may pose a potential challenge for machines to detect them. In addition, all the previous mentioned works used the whole storefront in their systems and faced some crucial issues in detection because of (1) limitations associated with existing datasets and (2) the indistinct boundaries of storefronts negatively influence the effective learning. This chapter aims to surmount these challenges by building a large-scale dataset, meticulously annotated where the focus is on signboards only. Subsequently, one-stage object detection models will be trained on this dataset to assess their performance.

6.2.2 *Storefront dataset*

Street View Text (SVT) dataset [17] is one of the earliest public datasets that focused on the signage of retail stores. It contains images of street views that include storefronts from Google Map.

These images had been harvested using Amazon's Mechanical Turk service.[a] The dataset has 350 images in which each image has one single store signboard. The total number of words in these images is 725. A set of 100 images with a total of 211 words was designated as the training set. The annotation for each image has been done by Alex Sorokin's Annotation Toolkit[b] which generates an xml format. For each image, business name and address are recorded in addition to image resolution, lexicon, and bounding boxes of words in the signage. The lexicon words have been collected from the top 20 business results that came out of searching all nearby business stores.

Another Dataset [1] was collected also using Google Street View with Google collaboration. Unfortunately, the dataset is not provided for the public even though the authors were contacted! The dataset contains 1.3 million street images with geo-information. They were collected from different urban areas in many cities across Europe, Australia, and the Americas. The annotation was done through some operators who generate bounding boxes around business-related information including signboards, however, the text was extracted using OCR software. In this study, the authors aimed to classify street view storefronts using their own set of 208 unique labels/classes.

Moreover, a Chinese shop sign dataset (ShopSign) [18] has been published for the public. It contains images of real street views for shop signboards. The images were collected by 40 students for more than two years using 50 different types of cameras. The ShopSign dataset has 25,770 Chinese shop sign images with geo-locations which include a total of 4,072 unique Chinese characters/classes. The annotation was done manually in a text-line-based manner with the help of 12 members using quadrilaterals. The dataset is large scale, and it is considered sparse and unbalanced. Moreover, the ShopSign dataset has big environmental and material diversity. Unfortunately, this dataset is based on Chinese characters with only limited samples of English scripts.

---

[a]http://mturk.com.
[b]http://vision.cs.uiuc.edu/annotation/.

6.3 The ShoS Dataset

6.3.1 *Data collection*

The **Shop Signboard Dataset ShoS** has been collected using Google Street View (GSV). This version of the dataset was collected from 51 cities in Canada and the USA including Toronto, Vancouver, Ottawa, Calgary, Edmonton, Chicago, Los Angeles, San Francisco, New York City, Seattle, Miami, and Boston. Using GSV, screenshot images of storefronts were taken after some adjustments to ensure clarity of the image. The recorded scene includes one or more shop signboards with a minimum of one store per image and a maximum of 7 stores per image (Figure 6.2). The average number of stores per image is 1.6 and more statistical descriptives are provided in Table 6.1. View angles were selected in a way that guarantees signboard visibility. The samples were collected by the researchers at first and by hired freelancers later through Upwork[c] freelance platform. Multiple revision cycles were completed to improve the quality of the collected data; this is further detailed in the following. The completed process of collecting and annotating the dataset took around one year.

The ShoS dataset contains 10k signboards within 7500 images of multiple resolutions but mainly in 3360×2100 and 1280×1024 pixels. It is available for public for multiple research purposes. Figure 6.1 shows some samples from the dataset.

6.3.2 *Data annotation*

The ShoS dataset was annotated using VGG Image Annotator (VIA) [19] which was developed by Visual Geometry Group at the University of Oxford. VIA is a manual annotation tool that is characterized by its simplicity and lightweight. It does not need any installation, and it works online with an interface or offline as an HTML file. VIA satisfied the requirements of this research as it enables adding more attributes to image files and bounding boxes where many tools do not. It also can generate annotation files as JSON and CSV files, however, additional modifications to the CSV file were needed to

---

[c]https://www.upwork.com/.

Fig. 6.1. Samples from the ShoS dataset illustrated with the bounding box annotations.

Table 6.1. Statistical descriptives of the number of signboard (region_count) per image.

Descriptives	region_count
N	10000
Missing	0
Mean	1.63
Median	1.00
Standard deviation	0.844
Variance	0.712

meet the research requirements. The final CSV file was then used to generate Pascal VOC XML and text files for each image through Python scripts.

For each image file, the following attributes were recorded: image name, image width, image height, and the number of bounding boxes in the image. Similarly, for each bounding box in each image, these

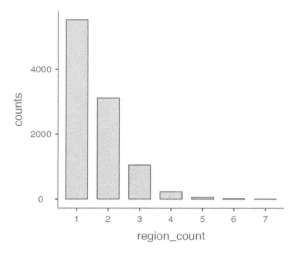

Fig. 6.2. Bar blot of the number of signboard (region_count) per image.

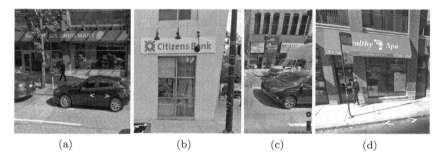

Fig. 6.3. Sample from the ShoS dataset of occluded signboards by (a) tree and shadow; (b) spotlight; (c) big vehicles; (d) traffic signs.

attributes were annotated: top left coordinates (xmin, ymin), bottom right coordinates (xmax, ymax), width, and height. Also, each bounding box was marked if it is occluded or not where shop signboards can be occluded by trees, tree branches, traffic signs, big vehicles, spotlights, shadow, and some other shop signs that are positioned for pedestrians. The signboard is considered occluded if at least 20% of the text on the sign is covered. About 9% of the ShoS signboards are occluded (see Figure 6.3 for examples of occluded signboards). In Figure 6.4, some samples from the CSV annotation file for the ShoS dataset are presented.

filename	image width	image height	region count	region id	xmin	ymin	width	height	xmax	ymax	occluded
img403.png	1280	823	2	0	469	242	211	114	680	356	no
img403.png	1280	823	2	1	776	249	176	127	952	376	no
img404.png	1280	823	1	0	660	229	450	69	1110	298	no
img405.png	3360	2100	2	0	831	580	966	251	1797	831	no
img405.png	3360	2100	2	1	1918	620	1175	206	3093	826	no
img406.png	1280	823	1	0	810	229	84	90	894	319	no

Fig. 6.4. Some samples from the CSV annotation file for the ShoS dataset.

Revising and cleaning the collected data proved to be a labor-intensive undertaking, necessitating several steps, particularly due to the diverse origins of the data obtained from freelancers. First, the hired freelancers were provided with a comprehensive guideline to follow during data collection. This included directives to avoid certain types of signboards, such as those made of painted glass, light-based signs, and board-less signs. An initial review was conducted to assess the collected images against the given criteria. Upon acceptance, the freelancers proceeded to annotate the images, following more detailed instructions.

To ensure data integrity, an annotation file was subjected to processing using a Python script. This step aimed to rectify null values and mismatched attributes. For instance, a street view image might be recorded as having three storefronts, while only two signboards were annotated. In such cases, individual investigation of the image was undertaken to address the discrepancy. This involved either annotating the missing signboard or rectifying the count of annotated regions. Figure 6.5 shows the workflow of the revising and cleaning step of the ShoS dataset.

6.3.3 *Challenges and limitation*

During data collection, multiple challenges were faced that limited the overall diversity of real world store signboards. The main challenges have been classified into three categories detailed in the following:

(A) Google Street View challenges: Some GSV factors impact the clarity of shop signboards. GSV collects street views by taking

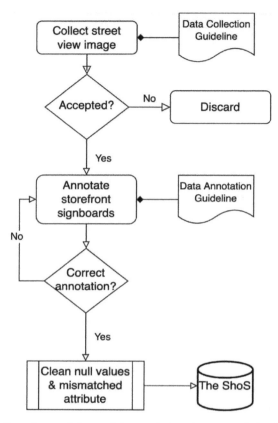

Fig. 6.5. A flow chart of the revising and cleaning step of the ShoS dataset.

panoramic view images and stitching them together [20]. Stitching images can create irregular shapes with repetitive or incomplete parts. Moreover, environmental factors may affect the clarity of GSV images, such as rain, which leads to foggy lens and thus foggy images. In addition, GSV images can be obstructed by traffic on the street, such as buses and large vehicles. All of these factors can obstruct shop signboards or impact the quality of their screenshots.

(B) Signboard material challenges: Different materials can be used to build shop signboards including mirror, glass, wood, or even bare wall. The study in [18] identified some of these material types as 'hard' for the detection and recognition process. Therefore, the number of signboards that are made of these material types was minimized from the ShoS dataset as much as possible.

(C) Signboard position challenges: Some shops place their signboards in abnormal positions. For example, shop signage can be perpendicular to the storefront in order to attract pedestrians. Also, it is possible to find shop signboards on top of buildings without an actual storefront. Capturing these types of signboards could confuse the learning process as they do not reflect the normal view of a storefront with a signboard. Thus, such images were eliminated from the ShoS dataset.

6.4 Experiments

For signboard detection, one-stage object detectors YOLOv3 [7] and SSD [8] were utilized. The models make several predictions in a single pass using convolutional neural networks. Several backbones, pre-trained base networks, were used based on the object detector where the chosen YOLO model uses Darknet53 pre-trained on the ImageNet dataset [21] and SSD model uses MobileNetv2 FPNLite pre-trained on COCO dataset [22]. Applying transfer learning helps boost performance by leveraging the knowledge gained from previous learning.

The ShoS dataset was preprocessed in order to train the YOLO and SSD models. All images underwent conversion from PNG to JPG format and were resized to dimensions of 960×720. This specific image size helped balance image quality and image processing time. Bigger image sizes could not be handled due to the excessive memory demands during training while smaller image sizes compromised model performance. Then, two distinct input resolutions were employed for training resized into 640×640 and 320×320. Additionally, two color schemes, RGB and grayscale, were incorporated. Thus, we would be able to investigate the potential impact of resolution and color factors on the accuracy of the trained models.

The dataset was partitioned into training and testing sets with an 80/20 ratio, respectively. Notably, this partitioning procedure was carefully executed to accommodate the variation in the number of stores depicted in each image. Given that certain images featured a singular store while others included multiple stores, a proportional representation from each group was ensured in both the training and test sets.

urthermore, a supplementary run was done with about 10% of the ShoS dataset random signboards which were intentionally blurred and included as true negative samples (Figure 6.6 presents a blurred sample). Remarkably, the outcomes of this modified approach exhibited a significant improvement in terms of mean average precision.

The number of classes c was set to one which is "Signboard". The number of batches was set as 64 and 16 for the smaller and bigger input resolutions, respectively, and the learning rate was 0.001. Some random augmentations were used including brightness, height shift, rescale, rotation, and zoom with respect to the overall structure of the image which is not destroyed in order to make the models robust to different variations as suggested in [23]. Figure 6.6 shows an illustrated sample for the signboard detector output during training.

The training was stopped after 5k iterations for YOLO and after 8k iterations for SSD as the results were plateauing. The output

Fig. 6.6. An illustrated sample for the signboard detector output during training.

of the inference stage was a bounding box for each signboard coupled with its confidence score. The confidence score C indicates how sure and accurate the model is regarding the bounding box containing a signboard. It was computed for each predicted signboard using the Jaccard index (AKA Intersection over Union (IoU)) and class probability using Eq. (6.1). Finally, the redundant overlapping bounding boxes with low confidence scores were eliminated using non-maximum suppression:

$$C = P(\text{object}) * \text{IoU}. \tag{6.1}$$

6.5 Results and Discussion

The experiments were performed using both models YOLO and SSD on Google Colab Pro[d] which provides access to a GPU machine with an option of "high-RAM" usage. Correct detection was defined if the intersection over union (IoU) value was 0.5 or higher, following recommendations from prior studies on license plate detection [10]. The models were tested for several variances of average precision over the four variations mentioned earlier encompassing input resolutions and color schemes.

The results, as depicted in Table 6.2, revealed that YOLOv3, with a higher input resolution of 640 × 640 and in the RGB color scheme, demonstrated superior performance, achieving a mean average precision of 94.23%. In comparison, SSD achieved an mAP of 90.4% at IoU = 0.5, with a confidence score set to 0.25 for both models. This indicates the significance of color appearance for enhanced detection accuracy. Additionally, it was observed that the spatial distribution of labels directly influences the detection of smaller signboards at lower resolutions, as they pose greater difficulty for detection. Figure 6.7 showcases sample results obtained by YOLOv3-640 and SSD-Mobilenet2-fpnlite-640 models highlighting the superiority of YOLOv3.

---

[d]https://colab.research.google.com.

Table 6.2. Results of one-stage detectors over mean average precision of 0.5 and 0.75 and the recall at IoU = 0.5.

Detector	Input Resolution	Image Color	%mAP @0.5	%mAP @0.75	%Recall
YOLOv3	320 × 320	RGB	92.0	66.09	91
	640 × 640	RGB	**94.23**	**76.76**	**94**
	320 × 320	Grayscale	91.2	64.09	89
	640 × 640	Grayscale	91.88	69.0	91
SSD-Moblilenet2	320 × 320	RGB	88.8	74.08	54
	640 × 640	RGB	90.4	76.84	53
	320 × 320	Grayscale	85.5	70.0	49
	640 × 640	Grayscale	84.6	68.8	47

Fig. 6.7. Results from YOLO3 640 showing high accuracy of detecting signboards even when they are occluded. Green bounding boxes indicate the ground truth and pink denotes all predications.

The robustness of the YOLOv3 model was evident in its ability to predict even partially occluded signboards. Notably, both detectors demonstrated strong capability to detect most signboards, including those intentionally missed by annotators due to aforementioned limitations. This allows a more accurate future stage for shops' classification based on their signboards [24]. Figures 6.8 and 6.9 present the performance charts for both methodologies. The average losses for all YOLOv3 and SSD variants were significantly decreasing during training reaching 0.07 and 0.02 respectively in their ultimate cases. The SSD model computes the loss differently when compared to YOLOv3 as it combines cross-entropy loss and

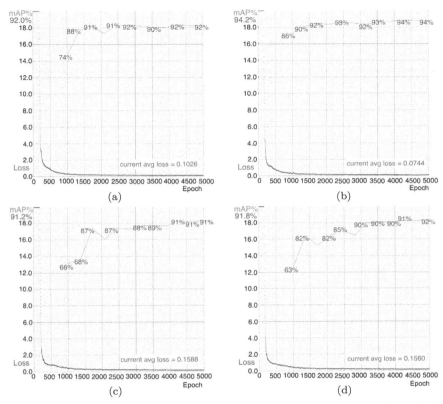

Fig. 6.8. Average loss (blue) and mean average precision mAP (red) for YOLO experiments: (a) RGB-320; (b) Grayscale-320; (c) RGB-640; (d) Grayscale-640.

Smooth L1 loss. However, it is observed that the recall of SSD variants, which assesses the model's ability to identify all true positive samples, was marginally lower in comparison to YOLOv3. This discrepancy may warrant further investigation in future research, particularly in the realm of Explainable AI (XAI), to elucidate the underlying reasons for such results.

Finally, it was observed that when we provided the detectors with true negative samples, i.e., blurred signboards, results were significantly higher by 10% approximately in terms of mean average precision. That negates the assumption made for one-stage detectors which ignores training the models with true negative samples. That was not the case at least in our research area.

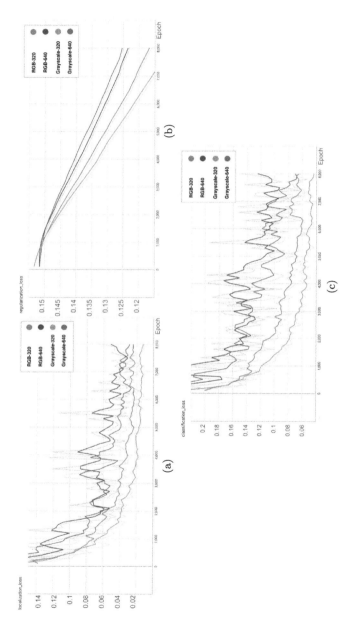

Fig. 6.9. (a) Localization, (b) regularization, and (c) classification loss charts for SSD experiments with a smooth factor set to 0.6.

6.6 Conclusion

This chapter makes a significant contribution by introducing the Shop Signboard (ShoS) dataset, complete with meticulous annotations for the purpose of shop signboard detection. A series of experiments were systematically conducted on the ShoS dataset utilizing one-stage object detection methods, and the ensuing results are comprehensively presented and discussed. This contribution paves the way for a more precise future stage involving the classification of shops based on their signboards and XAI to understand the behind scene superiority among models.

6.7 Dataset Availability

The full version of the ShoS dataset [24–26] is made public for research purposes. Interested parties may request access to the dataset by contacting CENPARMI research manager Nicola Nobile at nicola@cenparmi.concordia.ca or reaching out directly to the authors.

References

[1] Y. Movshovitz-Attias, Q. Yu, M. C. Stumpe, V. Shet, S. Arnoud and L. Yatziv, Ontological supervision for fine grained classification of street view storefronts, in *Proceedings of the IEEE Conference on Computer Vision and Pattern Recognition*, Boston, MA, USA, pp. 1693–1702 (2015).

[2] Q. Yu, C. Szegedy, M. C. Stumpe, L. Yatziv, V. Shet, J. Ibarz and S. Arnoud, Large scale business discovery from street level imagery (2015), *arXiv preprint* arXiv:1512.05430.

[3] S. S. Noorian, S. Qiu, A. Psyllidis, A. Bozzon and G.-J. Houben, Detecting, classifying, and mapping retail storefronts using street-level imagery, in *Proceedings of the 2020 International Conference on Multimedia Retrieval, ICMR'20*, Dublin, Ireland. Association for Computing Machinery, pp. 495–501 (2020) https://doi.org/10.1145/3372278.3390706.

[4] Building and Planning Department of the City of Westmount, Renovating and building in westmount — Storefronts and signage, https://www.westmount.org/wp-content/uploads/2014/07/7-Storefronts_and_Signage.pdf (2001). Accessed 30 March 2020.

[5] D. Erhan, C. Szegedy, A. Toshev and D. Anguelov, Scalable object detection using deep neural networks, in *Proceedings of the IEEE Conference on Computer Vision and Pattern Recognition*, Columbus, Ohio, USA, pp. 2147–2154 (2014).

[6] C. Szegedy, W. Liu, Y. Jia, P. Sermanet, S. Reed, D. Anguelov, D. Erhan, V. Vanhoucke and A. Rabinovich, Going deeper with convolutions, in *Proceedings of the IEEE Conference on Computer Vision and Pattern Recognition*, Boston, MA, USA, pp. 1–9 (2015).

[7] J. Redmon and A. Farhadi, Yolov3: An incremental improvement (2018), http://arxiv.org/abs/1804.02767.

[8] W. Liu, D. Anguelov, D. Erhan, C. Szegedy, S. Reed, C.-Y. Fu and A. C. Berg, SSD: Single shot multibox detector, in *European Conference on Computer Vision*, Amsterdam, Netherlands. Springer, pp. 21–37 (2016).

[9] S. Ren, K. He, R. Girshick and J. Sun, Faster r-cnn: Towards real-time object detection with region proposal networks, in C. Cortes, N. D. Lawrence, D. D. Lee, M. Sugiyama and R. Garnett (eds.), *Advances in Neural Information Processing Systems*, Vol. 28. Curran Associates, Inc., pp. 91–99 (2015), http://papers.nips.cc/paper/5638-faster-r-cnn-towards-real-time-object-detection-with-region-proposal-networks.pdf.

[10] R. Laroca, E. Severo, L. A. Zanlorensi, L. S. Oliveira, G. R. Gonçalves, W. R. Schwartz and D. Menotti, A robust real-time automatic license plate recognition based on the yolo detector, in *2018 International Joint Conference on Neural Networks (IJCNN)*, Rio, Brazil, pp. 1–10 (2018).

[11] J. Redmon, S. Divvala, R. Girshick and A. Farhadi, You only look once: Unified, real-time object detection, in *The IEEE Conference on Computer Vision and Pattern Recognition (CVPR)*, Las Vegas, NV, USA (2016).

[12] J. Redmon and A. Farhadi, Yolo9000: Better, faster, stronger, in *The IEEE Conference on Computer Vision and Pattern Recognition (CVPR)*, Honolulu, HI, USA (2017).

[13] R. Al-qudah and C. Y. Suen, Enhancing yolo deep networks for the detection of license plates in complex scenes, in *Proceedings of the Second International Conference on Data Science, E-Learning and Information Systems*, Dubai, UAE, pp. 1–6 (2019).

[14] J. Ren and H. Li, Implementation of vehicle and license plate detection on embedded platform, in *2020 12th International Conference on Measuring Technology and Mechatronics Automation (ICMTMA)*, Phuket, Thailand, pp. 75–79 (2020).

[15] A. Bochkovskiy, C.-Y. Wang and H.-Y. M. Liao, Yolov4: Optimal speed and accuracy of object detection (2020), https://arxiv.org/abs/2004.10934.

[16] G. Jocher, A. Stoken, J. Borovec *et al.*, ultralytics/yolov5: v3.1 — Bug Fixes and Performance Improvements (2020), https://doi.org/10.5281/zenodo.4154370.

[17] K. Wang and S. Belongie, Word spotting in the wild, in *European Conference on Computer Vision*. Springer, pp. 591–604 (2010).

[18] C. Zhang, G. Peng, Y. Tao, F. Fu, W. Jiang, G. Almpanidis and K. Chen, Shopsign: A diverse scene text dataset of chinese shop signs in street views (2019), *arXiv preprint* arXiv:1903.10412.

[19] A. Dutta and A. Zisserman, The {VIA} annotation software for images, audio and video **5** (2019), *arXiv preprint* arXiv:1904.10699.

[20] D. Anguelov, C. Dulong, D. Filip, C. Frueh, S. Lafon, R. Lyon, A. Ogale, L. Vincent and J. Weaver, Google street view: Capturing the world at street level, *Computer* **43**(6), pp. 32–38 (2010).

[21] O. Russakovsky, J. Deng, H. Su, J. Krause, S. Satheesh, S. Ma, Z. Huang, A. Karpathy, A. Khosla, M. Bernstein, A. C. Berg and L. Fei-Fei, ImageNet large scale visual recognition challenge, *International Journal of Computer Vision (IJCV)* **115**(3), pp. 211–252 (2015), doi:10.1007/s11263-015-0816-y.

[22] T.-Y. Lin, M. Maire, S. Belongie, L. Bourdev, R. Girshick, J. Hays, P. Perona, D. Ramanan, C. L. Zitnick and P. Dollár, Microsoft coco: Common objects in context (2015), arXiv:1405.0312 [cs.CV].

[23] X. Xie, X. Xu, L. Ma, G. Shi and P. Chen, On the study of predictors in single shot multibox detector, in *Proceedings of the International Conference on Video and Image Processing*, ICVIP 2017. ACM, New York, pp. 186–191 (2017), http://doi.acm.org.lib-ezproxy.concordia.ca/10.1145/3177404.3177412.

[24] M. Almuhajri and C. Y. Suen, A complete framework for shop signboards detection and classification, in *2022 26th International Conference on Pattern Recognition (ICPR)*. IEEE, pp. 4671–4677 (2022), doi:10.1109/ICPR56361.2022.9956399.

[25] M. Almuhajri and C. Y. Suen, Shop signboards detection using the shos dataset, in M. El Yacoubi, E. Granger, P. C. Yuen, U. Pal and N. Vincent (eds.), *Pattern Recognition and Artificial Intelligence*. Springer International Publishing, Cham, pp. 235–245 (2022).

[26] M. Almuhajri and C. Y. Suen, Ai based approach for shop classification and a comparative study with human, *Advances in Artificial Intelligence and Machine Learning AAIML* **2**(3), pp. 441–455 (2022), doi:10.54364/AAIML.2022.1129.

Chapter 7

Self-Distilled Self-Supervised Monocular Depth Estimation

Julio Mendoza and Helio Pedrini

*Institute of Computing, University of Campinas,
Campinas, SP, Brazil*

Abstract

This chapter explores methods for enhancing self-supervised monocular depth estimation models through self-distillation via prediction consistency. As some per-pixel depth predictions may be less accurate, we suggest a mechanism for filtering out unreliable predictions. Additionally, we examine various techniques for ensuring consistency between predictions. Our findings demonstrate that selecting appropriate filtering and consistency enforcement methods is critical for achieving significant improvements in monocular depth estimation. Our proposed approach delivers competitive performance on the KITTI benchmark.

7.1 Introduction

Depth estimation [1–4] is a fundamental task in computer vision, with applications ranging from 3D modeling to virtual and augmented reality and robot navigation. While sensors such as LIDAR or RGB-D cameras can provide depth information, their limited range and operating conditions make them unreliable in certain scenarios. This has made alternative approaches, such as estimating depth from images, more attractive. Supervised deep learning methods have shown impressive results in depth estimation. However, these approaches rely on costly acquisition of high-quality ground-truth depth data for training.

Self-supervised depth estimation methods differ from supervised methods in that they do not rely on ground-truth data. By using stereo images or monocular sequences as inputs, they can be trained on diverse datasets without depth labels. These approaches leverage geometric priors to learn image reconstruction as an auxiliary task, with depth maps being obtained as an intermediary result of the reconstruction process.

Numerous studies have demonstrated that self-supervised depth estimation models can be improved by learning additional auxiliary tasks, such as self-distillation. Self-distillation methods seek to enhance a model's performance by extracting knowledge from the model itself. One noteworthy technique for conducting self-distillation involves extracting information from distorted versions of input data, as proposed by Xu and Liu [5]. This is achieved by enforcing consistency between predictions generated from distorted versions of the same input.

This chapter presents a novel self-distillation approach to enhance self-supervised depth estimation in monocular videos by ensuring prediction consistency. To ensure that enforcing consistency between unreliable predictions does not yield useful knowledge, we propose a filtering strategy to remove such predictions.

Enforcing consistency between predictions has been a widely explored concept in both self-distillation [6–10] and semi-supervised learning [11–16]. To explore various strategies for enforcing consistency, we adapt and evaluate representative approaches on the task of self-supervised depth estimation.

Our proposed solution provides the following key contributions: (i) a novel multi-scale self-distillation method utilizing prediction consistency, (ii) the development of a filtering approach to remove unreliable per-pixel predictions from the pseudo-labels used in self-distillation, and (iii) an exploration and adaptation of various consistency enforcement strategies for self-distillation.

To validate the effectiveness of our proposed method, we conducted a thorough evaluation and compared its performance with state-of-the-art approaches on the KITTI benchmark. Our code is publicly available at https://github.com/jmendozais/SDSSDepth.

7.2 Related Work

In this section, we provide a brief overview of relevant literature related to the topics addressed in our work.

7.2.1 *Self-supervised depth estimation*

Self-supervised depth estimation methods leverage multi-view geometry relations calculated from depth and camera motion predictions to reconstruct one view with pixel values from another view. Depth and camera motion can be obtained from deep networks trained by minimizing reconstruction errors. Garg et al. [17] utilized this approach to train a depth network using stereo pairs as views. Similarly, Zhou et al. [18] proposed a method to extract views from monocular sequences and used deep networks to estimate relative pose and depth maps.

Overcoming limitations of self-supervised depth estimation methods has been a subject of interest in the literature. For instance, some methods have tackled issues, such as inaccurate predictions in occluded regions [19] or regions with moving objects [20, 21]. Additionally, other methods have attempted to enhance the learning signal by enforcing consistency between multiple representations of the scene [4, 19, 22] or using auxiliary tasks, such as semantic segmentation or self-distillation [23–25]. Similar to the approaches developed by Kaushik et al. [23] and Zhou et al. [25], our method employs self-distillation to improve depth estimation. However, unlike prior work, our approach focuses on exploring and adapting representative strategies to distill knowledge from our model's predictions.

7.2.2 *Pseudo-labeling approaches for self-supervised depth estimation*

Several self-supervised depth estimation methods use pseudo-labels to provide additional supervision for training their networks. Pseudo-labels can be obtained from classical stereo matching algorithms [26], external deep learning methods [27, 28], or the network's own predictions [29]. These methods have been trained from stereo images [26, 29] or monocular sequences [23, 24, 26–28].

Due to the unreliable quality of pseudo-labels, some methods adopt strategies to filter out unreliable per-pixel predictions. External confidence estimates [26–28] or uncertainty estimates obtained from the method itself [29] are used for this purpose.

In addition, there are methods that utilize multi-scale predictions to generate pseudo-labels. One approach is to use the highest-resolution prediction as a pseudo-label to supervise predictions at lower resolutions [30]. Another approach is to select the prediction with the lowest reconstruction error among the multi-scale predictions for each pixel to obtain the final pseudo-labels [31].

Our focus is on self-supervised methods that rely on their own predictions as pseudo-labels. For instance, Kaushik et al. [23] augmented a self-supervised method by performing a second forward pass with strongly perturbed inputs. The predictions from the second pass are supervised with predictions of the first pass. Liu et al. [24] proposed to leverage the observation that depth maps predicted from daytime images are more accurate than predictions from nighttime images. They used predictions from daytime images as pseudo-labels and trained a specialized network with nighttime images synthesized using a conditional generative model.

7.2.3 Self-distillation

These methods allow a model to leverage information from its own predictions to improve performance. One approach is to transfer knowledge from a previously trained instance of the model through predictions or features to a new instance of the model [6–10]. This process can be repeated iteratively. Self-distillation has a regularization effect on neural networks, reducing overfitting and increasing test accuracy at earlier iterations. However, after too many iterations, the model can underfit and test accuracy can decline [6].

Self-distillation has been widely studied in image classification problems, where methods often train multiple instances of a model sequentially, with the output of a model trained in a previous iteration used as a teacher for the model trained in the current iteration. For example, Furlanello et al. [7] proposed a method that performs distillation by training a series of models in sequence. Similarly, Yang et al. [8] introduced a technique that trains a model in a single training cycle to imitate the performance of multiple training cycles, using snapshots obtained at the end of the previous cycle as a teacher.

This chapter investigates the concept of utilizing multiple snapshots within a single training generation for self-supervised depth estimation.

7.2.4 *Consistency regularization*

Consistency regularization is a common approach used in deep semi-supervised learning to enforce consistency between predictions obtained from perturbed views of input examples.

The principle of enforcing consistency between predictions from perturbed views has been widely used in deep semi-supervised works through consistency regularization approaches. One of the early methods [11] used multiple forward passes on perturbed versions of input data. Later, it was shown that the use of strong data augmentation perturbations [15] or a combination of weak and strong data augmentation perturbations in a teacher–student training scheme [16] can further enhance the resulting models.

Previous studies have demonstrated that models whose weights are the average of the model being trained at different training steps, known as average models, can produce more accurate results in deep learning tasks [12–14]. These models can also be utilized as teachers to generate more precise pseudo-labels [12, 14]. Additionally, the implementation of cyclic learning rate schedulers can enhance the quality of the averaged models and ultimately improve the accuracy and generalization of the resulting model [13]. Furthermore, these methods can be adapted to the consistency regularization framework [14].

Our method employs a cyclic cosine annealing learning rate schedule, similar to the approach proposed by Athiwaratkun *et al.* [14], to improve the quality of the teacher model.

7.3 Proposed Method

This section outlines our approach to perform self-distillation through prediction consistency. We begin by introducing the necessary background on self-supervised depth estimation. Next, we present the key idea of our method, followed by a mechanism to eliminate unreliable per-pixel depth predictions. Finally, we provide an in-depth explanation of various techniques used to enforce prediction consistency.

7.3.1 *Preliminaries*

Our method is based on self-supervised depth estimation techniques that rely on view reconstruction as the primary source of supervision [18]. To achieve this, it is necessary to establish correspondences between pixel coordinates on frames that capture different views of the same scene. We denote these correspondences in Eq. (7.1):

$$x_s \sim \mathbf{K}\mathbf{T_{t \to s}}\mathbf{D_t}(x_t)\mathbf{K}^{-1}x_t. \tag{7.1}$$

The process of reconstructing the target frame involves using the correspondences and the pixel intensities in the source frame, represented as $\mathbf{\hat{I}s} \to \mathbf{t}(x_t) = \mathbf{Is}(x_s)$, through a technique known as image warping. This approach requires the dense depth map $\mathbf{D_t}$ of the target image, which we aim to reconstruct, the Euclidean transformation $\mathbf{T_{t \to s}}$, and camera intrinsics $\mathbf{K}$. To predict the depth maps and the Euclidean transformation, we utilize a convolutional neural network and assume that the camera intrinsics are given. The networks are trained using the adaptive consistency loss $\mathcal{L}_{\text{ac}}$, and further details can be found in [4].

7.3.2 *Self-distillation via prediction consistency*

The main concept behind self-distillation based on prediction consistency is to enhance the supervision provided to the model by ensuring that the predicted depth maps from various perturbed views of an input image are consistent with each other.

Our self-distillation method leverages two distinct data augmentation perturbations applied to an input snippet. To reduce the computational burden, we consider snippets consisting of only two frames, denoted as $\mathcal{I} = \mathbf{I_t}, \mathbf{I_{t+1}}$. The model predicts the depth maps for both frames in the snippet. As we use two different data augmentation perturbations, we obtain two sets of predicted depth maps for each frame. To enforce consistency, we minimize the difference between the predicted depth maps for each frame.

There are various ways to enforce consistency between predictions. The most straightforward variation of our approach uses the pseudo-label method. This involves designating one of the depth maps as the pseudo-label $\mathbf{D}^{(\text{pl})}$, which means that gradients are not back-propagated through it, and the other depth map as the prediction $\mathbf{D}^{(\text{pred})}$. In Section 7.3.4, we enhance our method by exploring other strategies for enforcing consistency.

Furthermore, our method utilizes the mean squared error (MSE) as the difference measure for enforcing prediction consistency. Additionally, we apply a composite mask to filter out unreliable depth values on the pseudo-label. The self-distillation loss term for a snippet $\mathcal{I}$ is presented in Eq. (7.2):

$$\mathcal{L}_{\text{sd}} = \frac{1}{|\mathcal{I}|} \sum_{\mathbf{I}_k \in \mathcal{I}} \frac{1}{|\mathbf{M}_k^{(\text{c})}|} \sum_{x \in \Omega(\mathbf{I}_k)} \mathbf{M}_k^{(\text{c})}(x) \left(\mathbf{D}_k^{(\text{pl})}(x) - \mathbf{D}_k^{(\text{pred})}(x) \right)^2, \tag{7.2}$$

where $\mathbf{I}_k$ represents a frame in the snippet, $\mathbf{M}_k^{(\text{c})}$ is its composite mask, x denotes a pixel coordinate, $\Omega(\mathbf{I}_k)$ represents the set of pixel coordinates, and $\mathbf{D}_k^{(\text{pl})}$ and $\mathbf{D}_k^{(\text{pred})}$ represent the pseudo-label and predicted depth maps, respectively.

Given that our model predicts depth maps at multiple scales, we apply the self-distillation loss at each scale. We assume that the pseudo-label at the finest scale is the most accurate, and therefore, we only use this pseudo-label. To match the scale of the pseudo-label, we upsample the predictions to the finest scale. Finally, we compute the self-distillation loss for each scale using the MSE as a difference measure. Figure 7.1 provides an overview of our self-distillation approach.

7.3.3 *Filtering pseudo-labels*

We observed through experimentation that when the depth predictions are unreliable, there are very large differences between the pseudo-labels and predictions. These differences can lead to instability during training and can prevent the model from converging. To address this issue, we exclude pixels with very large differences by applying a threshold value. In this section, we describe two methods for determining the threshold.

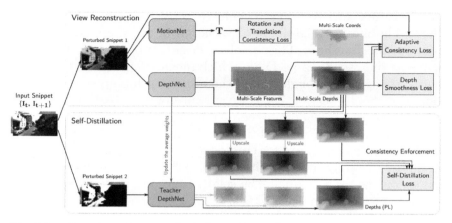

Fig. 7.1. Overview of our method. Our self-distillation method builds upon the multi-scale depth map predictions obtained from the view reconstruction component. These predictions are upscaled to the finest resolution and the teacher model's more accurate predictions are used as pseudo-labels to enhance the predictions obtained from view reconstruction.

One approach to determining the threshold is to compute it as a percentile P based on the differences between the pseudo-labels and predictions for all pixels in a batch of snippets. Then, we create a valid mask by considering all pixels with differences smaller than the threshold as valid, as shown in Eq. (7.3):

$$\mathbf{M}^{(p)}(x) = \left[\left(\mathbf{D}^{(\mathbf{pl})}(x) - \mathbf{D}^{(\mathbf{pred})}(x)\right)^2 < P\right], \qquad (7.3)$$

where [.] denotes the Iverson bracket operator. The final mask is obtained by combining the latter mask with the compound mask. The final mask could be expressed as $\mathbf{M} = \mathbf{M}^{(p)} \odot \mathbf{M}^{(c)}$, where $\odot$ represents the element-wise product. Finally, we replace $\mathbf{M}^{(c)}$ with $\mathbf{M}$ in Eq. (7.2).

We hypothesize that the approach of using a threshold based on the distribution of differences across batches may not be ideal since it does not account for the fact that batches with more accurate predictions should have a higher threshold that excludes fewer pixels than batches with less reliable predictions.

Our second scheme addresses the limitation of the first one by approximating a global threshold $P^{(\mathrm{EMA})}$ using the exponential moving average (EMA) of the percentile values for each batch during training. This approach has the advantage of taking into consideration that the distribution of depth differences changes during training. As the depth differences become smaller, the threshold increases by giving more weight to the percentiles from later batches in the average. Equation (7.4) presents our method for approximating the global threshold:

$$P_t^{(\text{EMA})} = P_{t-1}^{(\text{EMA})} \cdot \beta + P_t \cdot (1 - \beta), \tag{7.4}$$

where P_t is the threshold computed from the batch at the t training iteration, $P_t^{(\text{EMA})}$ is the threshold obtained using the EMA at the t training iteration, and β controls the influence of the previous moving average percentile and the current percentile into the computation of the current threshold. Similar to the first scheme, we compute a valid mask $\mathbf{M}^{(\text{EMA})}$ using $P^{(\text{EMA})}$, we combine this mask with the compound mask $\mathbf{M} = \mathbf{M}^{(\text{EMA})} \odot \mathbf{M}^{(c)}$, and, finally, we use $\mathbf{M}$ instead of $\mathbf{M}^{(c)}$ in Eq. (7.2).

7.3.4 *Consistency enforcement strategies*

In the preceding sections, we discussed a technique of pseudo-labeling to ensure the coherence between depth forecasts. In this section, we present typical consistency enforcement techniques adjusted to our self-distillation method.

These consistency enforcement strategies are illustrated in Figure 7.2 and have been adapted to our self-distillation approach. We have named each strategy similar to the methods that introduced the key idea in the semi-supervised learning literature [11–14].

Like the pseudo-label approach, the methods presented in this section also employ the second scheme discussed in Section 7.3.3 to eliminate unreliable per-pixel predictions before calculating the prediction differences.

7.3.4.1 Π-*Model*

The strategy discussed here is similar to the pseudo-label approach, as it also aims to enforce consistency between depth predictions. However, instead of using a pseudo-label, this approach enforces consistency between depth predictions from two perturbed views of the

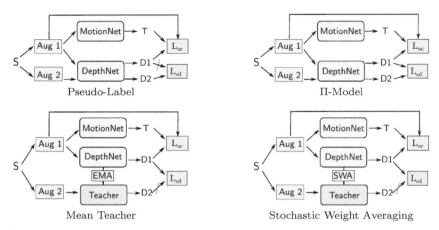

Fig. 7.2. Simplified views of the consistency enforcement strategies. S denotes the input snippet, $Aug\ 1$ and $Aug\ 2$ denote two perturbed views of the input snippet, T denotes the camera motion transformation, $D1$ and $D2$ denote depth maps predictions, L_{ac} denotes the adaptive consistency loss, L_{sd} denotes the self-distillation loss, and red lines —— mark connections where the gradients are not back-propagated.

same input. Unlike the pseudo-label approach, the gradients are back-propagated through both predictions, allowing for better alignment of the features used for prediction.

7.3.4.2 Mean Teacher

Instead of relying on the same depth network to generate both pseudo-labels and predictions, an alternative approach is to introduce a teacher network that can potentially produce more accurate pseudo-labels and provide better supervision for the student network being trained. In this approach, the weights of the teacher depth network are determined as the exponential moving average (EMA) of the depth network weights at equally spaced training iterations.

7.3.4.3 Stochastic Weight Averaging

The cyclic mean teacher strategy also involves setting the teacher depth network weights as the EMA of the depth network weights. However, the training process is divided into multiple cycles, with the learning rate decreasing at each cycle. At the end of each training cycle, the teacher depth network is updated with the weights of the depth network from the last epoch, where the learning rate reaches its minimum value.

During the first generation of the training process, we use the student network to generate pseudo-labels. After the model has converged to a satisfactory local optimum, we initialize the teacher network weights with the student network weights. In the following training cycles, we simulate multiple generations of training by using a cyclic cosine annealing learning rate. At the end of each cycle, when the learning rate reaches its minimum value and the model is likely to have converged to a good local optimum, we update the teacher network weights using the exponential moving average (EMA) with the student network weights.

7.3.5 Additional considerations

Final Loss: The final loss is a combination of several loss terms, including our self-distillation loss $\mathcal{L}$sd, adaptive consistency loss $\mathcal{L}$ac [4], depth smoothness loss $\mathcal{L}$ds, translation consistency loss $\mathcal{L}$tc, and rotation consistency loss $\mathcal{L}_{\text{rc}}$. Our translation consistency loss only takes camera motion into consideration, while the rotation and translation consistency losses are similar to the cyclic consistency loss defined by Gordon et al. [20]. The final loss is calculated as a weighted sum of all these loss terms and is expressed in Eq. (7.5):

$$\mathcal{L} = \sum_{i \in \mathcal{S}} \frac{1}{2^i} \left(\mathcal{L}_{\text{ac}}^{(i)} + \lambda_{\text{ds}} \mathcal{L}_{\text{ds}}^{(i)} + \lambda_{\text{sd}} \mathcal{L}_{\text{sd}}^{(i)} \right) + \lambda_{\text{rc}} \mathcal{L}_{\text{rc}} + \lambda_{\text{tc}} \mathcal{L}_{\text{tc}}, \qquad (7.5)$$

where $\mathcal{S}$ is the set of scales and λ_{sd}, λ_{ds} λ_{rc}, λ_{tc} are the weight of the self-distillation, depth smoothness, rotation consistency, and translation consistency loss terms, respectively.

Network Architecture: We use a convolutional encoder–decoder network with skip connections as our depth network. Our encoder uses a ResNet18 architecture, and our decoder stacks convolutional and up-sampling layers. The intermediate layers in the convolutional layers use the ELU activation function [32], while the output layers use the sigmoid activation function. In contrast to the depth networks described by Godard *et al.* [33] that predict disparity, we predict depth using our network. This change has reduced artifacts with very high depth values in our outputs.

The motion network we use also has a ResNet18 backbone for feature extraction but with a modification to allow for multi-frame inputs in the form of snippets. The head of the motion network consists of four convolutional layers, which leverage the spatial information learned by the feature extractor. In the last layer of the head part of the motion network, we use global average pooling.

For more information about the network architecture, we refer the readers to our source code, which is publicly available at https://github.com/jmendozais/SDSSDepth.

7.4 Evaluation

Our experimental evaluation aims to address the following questions: (1) Does incorporating a multi-scale self-distillation via prediction consistency loss term improve the performance of a self-supervised trained model based on view reconstruction? (2) Do our methods to filter out unreliable per-pixel predictions enhance the performance of a model trained with self-supervised and self-distillation loss terms? (3) Which consistency enforcement strategies are the most effective when used with our multi-scale self-distillation via prediction consistency loss term?

To address the first question, we conduct a comparison between a competitive baseline and a modified version of the same baseline that integrates our multi-scale self-distillation approach based on prediction consistency, as outlined in Section 7.4.2. To answer the second question, we compare the performance of the simplest version of our self-distillation method with two alternative versions that incorporate our per-pixel filtering techniques described in Section 7.4.3. To address the third question, we compare the effectiveness of the various consistency enforcement strategies employed in our multi-scale self-distillation approach, as discussed in Section 7.4.4. Additionally, we present qualitative results and a comparative analysis with state-of-the-art approaches evaluated in similar scenarios.

7.4.1 Experimental setup

In this section, we provide details about the dataset used for training and evaluation as well as the model parameters and optimization method used to train the main variants of our proposed method.

Dataset

We utilized the KITTI benchmark [34], which consists of video sequences comprising 93,000 images captured using high-quality RGB cameras while driving on rural roads and highways in a city.

We used the Eigen split [35], which consists of 45023 images for training and 687 images for testing from the KITTI dataset. We further divided the training set into 40441 samples for training and 4582 samples for validation. Additionally, we evaluated our results using the same metrics as in the Eigen evaluation and clipped the predicted depth to 80 m.

Training

Our optimization algorithm of choice is ADAM with hyperparameters β_1 set to 0.9 and β_2 set to 0.999. We used a batch size of four snippets, where each snippet contains two frames unless stated otherwise. We resized the frames to 416×128 pixels, unless specified otherwise.

Our training process consists of several stages. Initially, we train our models without the self-distillation loss. During this stage, we use a learning rate of $1e-4$ for 15 epochs, followed by a reduction to $1e-5$ for another 10 epochs.

In the second stage of our training process, we train all models that include a self-distillation term using a learning rate of $1e-5$ for 10 epochs. Finally, in the third stage, we train the variants of our model that use teacher networks with average weights.

The mean teacher model is trained for 10 epochs in the third stage, with weight updates every $1e3$ iterations. In contrast, for the SWA model, we use a cyclical cosine learning rate schedule with an upper bound of $1e-4$, a lower bound of $1e-5$, and four cycles of six epochs each during this stage.

Table 7.1. Comparison of the baseline model and the variation of our method that uses the pseudo-labeling strategy.

Method	↓ Lower is better				↑ Higher is better		
	Abs Rel	Sq Rel	RMSE	LRMSE	$\delta < 1.25$	$\delta < 1.25^2$	$\delta < 1.25^3$
Baseline	0.128	1.005	5.152	0.204	**0.848**	**0.951**	0.979
PL	**0.126**	**0.907**	**5.068**	**0.202**	0.847	**0.951**	**0.980**

7.4.2 Self-distillation via prediction consistency

The comparison between the baseline and the simplest variant of our self-distillation method is shown in Table 7.1. The baseline, which is trained with two-frame snippets, achieves a performance that is competitive with the widely used baseline [33] that uses three-frame snippets for training. The results demonstrate a consistent improvement when the self-distillation loss is incorporated. The model trained with the self-distillation loss outperforms the baseline in terms of all error metrics and almost all accuracy metrics.

During our search for the optimal weight λ_{sd} for the self-distillation term, we observed that larger λ_{sd} values can lead to better results. However, we also noticed that using a high value of λ_{sd} can cause the model to diverge in some cases due to large depth differences. Therefore, we opted for a smaller value of $\lambda_{sd} = 1e2$. This instability in the model prompted us to investigate methods for filtering out unreliable predictions.

7.4.3 Filtering pseudo-labels

Table 7.2 demonstrates that both of our filtering strategies outperform the variation of our method that does not utilize any additional filtering approach apart from the composite mask in the majority of error and accuracy metrics. Furthermore, the results illustrate that the approach that leverages the exponential moving average (EMA) of the percentiles to estimate the threshold performs better than the approach using only the percentile of each batch.

Table 7.2. Comparison of variants of our method with and without filtering strategies.

Method	↓ Lower is better				↑ Higher is better		
	Abs Rel	Sq Rel	RMSE	LRMSE	$\delta < 1.25$	$\delta < 1.25^2$	$\delta < 1.25^3$
PL (w/o filtering)	0.126	0.907	5.068	0.202	0.847	0.951	0.980
PL + P	0.126	0.911	5.033	0.203	0.847	0.952	0.980
PL + $P^{(\text{EMA})}$	0.126	0.904	5.024	0.202	0.847	0.952	0.980

Notes: The notation P indicates that we used the percentile by batch as thresholds for filtering pseudo-labels, while $P^{(\text{EMA})}$ indicates that we used a threshold that is the EMA of the percentiles computed from the batches during training iterations.

Table 7.3. Comparison of different consistency enforcement strategies.

Method	↓ Lower is better				↑ Higher is better		
	Abs Rel	Sq Rel	RMSE	LRMSE	$\delta < 1.25$	$\delta < 1.25^2$	$\delta < 1.25^3$
Baseline	0.128	1.005	5.152	0.204	**0.848**	0.951	0.979
PL + $P^{(\text{EMA})}$	0.126	0.904	**5.024**	0.202	0.847	**0.952**	0.980
Π-M + $P^{(\text{EMA})}$	0.126	0.902	5.041	0.202	0.847	**0.952**	0.980
MT + $P^{(\text{EMA})}$	0.126	0.898	5.061	**0.201**	0.846	**0.952**	**0.981**
SWA + $P^{(\text{EMA})}$	**0.125**	**0.881**	5.056	0.202	**0.848**	**0.952**	0.980

Notes: PL stands for Pseudo-Label, Π M stands for Π-Model, MT stands for Mean Teacher, and SWA stands for Stochastic Weight Averaging.

7.4.4 Consistency enforcement strategies

Table 7.3 presents the results of our experiments comparing different consistency enforcement strategies in our self-distillation approach. Our findings suggest that, irrespective of the strategy used, self-distillation via prediction consistency enhances the performance of our baseline model. Furthermore, the results demonstrate that the variant employing SWA strategy outperforms the other approaches in the majority of error and accuracy metrics. Consequently, this variant is chosen as our final model.

Figure 7.3 presents qualitative results of our final model. The predicted depth maps are visually pleasing, with sharp edges on salient objects of the image. However, we observe that our model fails to predict a consistent depth map for a thin object with a variable background, as shown in the bottom-right image.

Fig. 7.3. Qualitative results. Depths maps obtained using our final model.

7.4.5 *State-of-the-art comparison*

Table 7.4 presents a quantitative comparison between our approach and state-of-the-art methods. Our method surpasses techniques that specifically address moving objects, such as [20–22]. Moreover, our method obtains competitive results compared to the state-of-the-art methods in terms of accuracy and error metrics.

7.5 Conclusions

Our study highlights the need for additional strategies to fully benefit from self-distillation in self-supervised depth estimation from monocular videos. One such strategy we considered was filtering out unreliable per-pixel predictions using a threshold value.

Furthermore, we illustrated the significance of selecting an appropriate consistency enforcement strategy in self-distillation. Our findings indicate that the SWA consistency enforcement strategy is crucial for achieving greater improvements, as it reinforces teacher quality and emphasizes the differences between teacher and student network weights.

Table 7.4. Comparison with the state of the art on the Eigen split of the KITTI dataset. We compared our results against several methods of the literature. To allow a fair comparison, we report the results of competitive methods trained with a resolution of 416×128 pixels.

Method	N.F.	↓ Lower is better				↑ Higher is better		
		Abs Rel	Sq Rel	RMSE	LRMSE	$\delta < 1.25$	$\delta < 1.25^2$	$\delta < 1.25^3$
Gordon et al. [20]	2	0.129	0.959	5.230	0.213	0.840	0.945	0.976
Our method	2	**0.125**	**0.881**	**5.056**	**0.202**	**0.848**	**0.952**	**0.980**
Zhou et al. [18]*	3	0.183	1.595	6.709	0.270	0.734	0.902	0.959
Mahjourian et al. [19]	3	0.163	1.240	6.220	0.250	0.762	0.916	0.967
Casser et al. [21]	3	0.141	1.026	5.290	0.215	0.816	0.945	0.979
Chen et al. [22]	3	0.135	1.070	5.230	0.210	0.841	0.948	0.980
Godard et al. [33]	3	0.128	1.087	5.171	0.204	0.855	0.953	0.978
Our method	3	0.123	**0.906**	5.083	0.200	0.856	0.953	0.980
Fang [36]	3	**0.116**	–	**4.850**	**0.192**	**0.871**	**0.959**	**0.982**

Notes: N.F. denotes the number of frames in the input snippet and (*) indicates new results obtained from an official repository. (-ref.) indicates that the online refinement component is disabled.

The improvements achieved with the different versions of our method are consistent with the discoveries of recent studies that have incorporated self-distillation in their loss function [23, 24]. Furthermore, the mechanisms employed in our method are fully compatible with these works [23, 24] and can be seamlessly integrated.

We believe that our research could offer valuable insights for utilizing self-distillation in stereo-based methods as well as supervised and semi-supervised methods.

References

[1] A. Lopes, R. Souza and H. Pedrini, A survey on RGB-D datasets, *Computer Vision and Image Understanding* **222**, pp. 1–22 (2022).

[2] J. Mendoza and H. Pedrini, Self-supervised depth estimation based on feature sharing and consistency constraints, in *15th International Conference on Computer Vision Theory and Applications*, Valletta, Malta, pp. 134–141 (2020).

[3] A. Pinto, M. Cordova, L. Decker, J. Flores-Campana, M. Souza, A. Santos, J. Conceiçao, H. Gagliardi, D. Luvizon, R. Torres and H. Pedrini, Parallax motion effect generation through instance segmentation and depth estimation, in *IEEE International Conference on Image Processing*, Abu Dhabi, United Arab Emirates, pp. 1621–1625 (2020).

[4] J. Mendoza and H. Pedrini, Adaptive self-supervised depth estimation in monocular videos, in *International Conference on Image and Graphics*, Haikou, China. Springer, pp. 687–699 (2021).

[5] T.-B. Xu and C.-L. Liu, Data-distortion guided self-distillation for deep neural networks, in *AAAI Conference on Artificial Intelligence*, Vol. 33, pp. 5565–5572 (2019).

[6] H. Mobahi, M. Farajtabar and P. Bartlett, Self-distillation amplifies regularization in Hilbert space, in H. Larochelle, M. Ranzato, R. Hadsell, M. F. Balcan and H. Lin (eds.), *Advances in Neural Information Processing Systems*, Vol. 33. Curran Associates, Inc., pp. 3351–3361 (2020).

[7] T. Furlanello, Z. Lipton, M. Tschannen, L. Itti and A. Anandkumar, Born again neural networks, in *International Conference on Machine Learning*. PMLR, pp. 1607–1616 (2018).

[8] C. Yang, L. Xie, C. Su and A. L. Yuille, Snapshot distillation: Teacher-student optimization in one generation, in *IEEE/CVF Conference on Computer Vision and Pattern Recognition*, pp. 2859–2868 (2019).

[9] Z. Zhang and M. R. Sabuncu, Self-distillation as instance-specific label smoothing (2020), *arXiv preprint* arXiv:2006.05065.

[10] K. Kim, B. Ji, D. Yoon and S. Hwang, Self-knowledge distillation with progressive refinement of targets, in *IEEE/CVF International Conference on Computer Vision*, pp. 6567–6576 (2021).

[11] M. Sajjadi, M. Javanmardi and T. Tasdizen, Regularization with stochastic transformations and perturbations for deep semi-supervised learning, *Advances in Neural Information Processing Systems*, Vol. 29, pp. 1163–1171 (2016).

[12] A. Tarvainen and H. Valpola, Mean teachers are better role models: Weight-averaged consistency targets improve semi-supervised deep learning results, in I. Guyon, U. V. Luxburg, S. Bengio, H. Wallach, R. Fergus, S. Vishwanathan and R. Garnett (eds.), *Advances in Neural Information Processing Systems*, Vol. 30. Curran Associates, Inc., pp. 1–10 (2017).

[13] P. Izmailov, D. Podoprikhin, T. Garipov, D. Vetrov and A. G. Wilson, Averaging weights leads to wider optima and better generalization (2018), *arXiv preprint* arXiv:1803.05407.

[14] B. Athiwaratkun, M. Finzi, P. Izmailov and A. G. Wilson, There are many consistent explanations of unlabeled data: why you should average, *International Conference on Learning Representations* (2019).

[15] Q. Xie, Z. Dai, E. Hovy, T. Luong and Q. Le, Unsupervised data augmentation for consistency training, *Advances in Neural Information Processing Systems*, **33** (2020).

[16] K. Sohn, D. Berthelot, N. Carlini, Z. Zhang, H. Zhang, C. A. Raffel, E. D. Cubuk, A. Kurakin and C.-L. Li, FixMatch: Simplifying semi-supervised learning with consistency and confidence, *Advances in Neural Information Processing Systems*, **33** (2020).

[17] R. Garg, V. K. BG, G. Carneiro and I. Reid, Unsupervised CNN for single view depth estimation: Geometry to the rescue, in *European Conference on Computer Vision*. Springer, pp. 740–756 (2016).

[18] T. Zhou, M. Brown, N. Snavely and D. G. Lowe, Unsupervised learning of depth and ego-motion from video, in *IEEE Conference on Computer Vision and Pattern Recognition*, pp. 1851–1858 (2017).

[19] R. Mahjourian, M. Wicke and A. Angelova, Unsupervised learning of depth and ego-motion from monocular video using 3D geometric constraints, in *IEEE Conference on Computer Vision and Pattern Recognition*, pp. 5667–5675 (2018).

[20] A. Gordon, H. Li, R. Jonschkowski and A. Angelova, Depth from videos in the wild: Unsupervised monocular depth learning from unknown cameras (2019), *arXiv preprint* arXiv:1904.04998.

[21] V. Casser, S. Pirk, R. Mahjourian and A. Angelova, Depth prediction without the sensors: Leveraging structure for unsupervised learning from monocular videos, in *AAAI Conference on Artificial Intelligence*, Vol. 33, pp. 8001–8008 (2019).

[22] Y. Chen, C. Schmid and C. Sminchisescu, Self-supervised learning with geometric constraints in monocular video: connecting flow, depth, and camera, in *IEEE International Conference on Computer Vision*, pp. 7063–7072 (2019).

[23] V. Kaushik, K. Jindgar and B. Lall, ADAADepth: Adapting data augmentation and attention for self-supervised monocular depth estimation (2021), *arXiv preprint* arXiv:2103.00853.

[24] L. Liu, X. Song, M. Wang, Y. Liu and L. Zhang, Self-supervised monocular depth estimation for all day images using domain separation, in *IEEE/CVF International Conference on Computer Vision*, pp. 12737–12746 (2021).

[25] H. Zhou, S. Taylor and D. Greenwood, SUB-depth: Self-distillation and uncertainty boosting self-supervised monocular depth estimation (2021), *arXiv e-prints* arXiv–2111.

[26] A. Tonioni, M. Poggi, S. Mattoccia and L. Di Stefano, Unsupervised domain adaptation for depth prediction from images, *IEEE Transactions on Pattern Analysis and Machine Intelligence* **42**(10), pp. 2396–2409 (2019).

[27] H. Choi, H. Lee, S. Kim, S. Kim, S. Kim, K. Sohn and D. Min, Adaptive confidence thresholding for monocular depth estimation, in *IEEE/CVF International Conference on Computer Vision*, pp. 12808–12818 (2021).

[28] J. Cho, D. Min, Y. Kim and K. Sohn, Deep monocular depth estimation leveraging a large-scale outdoor stereo dataset, *Expert Systems with Applications* **178**, p. 114877 (2021).

[29] H. Xu, Z. Zhou, Y. Wang, W. Kang, B. Sun, H. Li and Y. Qiao, Digging into uncertainty in self-supervised multi-view stereo, in *IEEE/CVF International Conference on Computer Vision*, pp. 6078–6087 (2021).

[30] J. Yang, J. M. Alvarez and M. Liu, Self-supervised learning of depth inference for multi-view stereo, in *IEEE/CVF Conference on Computer Vision and Pattern Recognition*, pp. 7526–7534 (2021).

[31] R. Peng, R. Wang, Y. Lai, L. Tang and Y. Cai, Excavating the potential capacity of self-supervised monocular depth estimation, in *IEEE International Conference on Computer Vision*, pp. 15560–15569 (2021).

[32] D.-A. Clevert, T. Unterthiner and S. Hochreiter, Fast and accurate deep network learning by exponential linear units (ELUs) (2015), *arXiv preprint* arXiv:1511.07289.

[33] C. Godard, O. Mac Aodha, M. Firman and G. J. Brostow, Digging into self-supervised monocular depth prediction, in *International Conference on Computer Vision*, pp. 3828–3838 (2019).

[34] A. Geiger, P. Lenz, C. Stiller and R. Urtasun, Vision meets robotics: The KITTI dataset, *The International Journal of Robotics Research* **32**(11), pp. 1231–1237 (2013).

[35] D. Eigen, C. Puhrsch and R. Fergus, Depth map prediction from a single image using a multi-scale deep network, in *Advances in Neural Information Processing Systems*, pp. 2366–2374 (2014).

[36] Z. Fang, X. Chen, Y. Chen and L. V. Gool, Towards good practice for cnn-based monocular depth estimation, in *IEEE/CVF Winter Conference on Applications of Computer Vision*, pp. 1091–1100 (2020).

Chapter 8

An Encoder–Decoder Approach to Offline Handwritten Mathematical Expression Recognition with Residual Attention

Qiqiang Lin*, Chunyi Wang*, Ning Bi*, Ching Y. Suen[†], and Jun Tan*

*School of Mathematics, Sun Yat-Sen University, China
[†]Centre for Pattern Recognition and Machine Intelligence, Concordia University, Canada

Abstract
The process of recognizing corresponding LaTeX sequences from pictures of handwritten mathematical expressions (MEs) is called handwritten mathematical expression recognition (HMER). With the rapid technological advancements in contemporary society, the accurate recognition of handwritten MEs has emerged as a significant research challenge. Unlike handwritten numeral recognition and handwritten text recognition, the recognition of handwritten mathematical expressions is more difficult because of their complex two-dimensional spatial structure. Since the "watch, attend and parse (WAP)" method was proposed in 2017, encoder–decoder models have made significant progress in handwritten mathematical expression recognition. Our model is improved based on the WAP model. In this chapter, the attention module is reasonably added to the encoder to make the extracted features more informative. The new network is called Dense Attention Network (DATNet), which allows for an adequate extraction of the structural information from handwritten mathematical expressions. To prevent

the model from overfitting during the training process, we use label smoothing as the loss. Experiments showed that our model (DATWAP) improved WAP expression recognition from 48.4%/46.8%/45.1% to 54.72%/52.83%/52.54% on the CROHME 2014/2016/2019 test sets.

8.1 Introduction

8.1.1 *Handwritten mathematical expression recognition*

There are currently two main types of systems for recognizing handwritten mathematical expressions: online recognition systems and offline recognition systems. The primary difference between these two systems lies in the input data they use. The data for online recognition systems are sequences of handwriting points sampled in chronological order, while the data for offline recognition systems are static images, which are typically scanned images of paper documents. This loss of dynamic trajectory information makes recognition more challenging. This chapter focuses specifically on the offline recognition system.

Handwritten mathematical expression recognition (HMER) is more complicated than recognizing one-dimensional handwriting (handwritten numbers and hand-written text). Since one-dimensional handwriting often has the same symbol size and the symbols are clearly spaced apart, symbol segmentation and recognition is easier. Mathematical expressions, on the other hand, are two-dimensional structures made up of mathematical symbols. To recognize mathematical expressions correctly, not only do the mathematical symbols contained within them need to be recognized accurately, but also the two-dimensional relationships between the symbols need to be resolved accurately, any tiny errors in these two processes will lead to a failure. Recognizing mathematical expressions is a challenging task due to various difficulties that arise during the recognition process. These difficulties have been identified in previous studies [1, 2] and include the following:

- **Context-sensitive symbol problems:** Mathematical symbols can have different meanings depending on their context.
- **2D structure parsing problems:** Mathematical expressions are composed of complex 2D structures, with a hierarchical

relationship between the various symbols and sub-expressions. This structure needs to be accurately parsed and understood to correctly recognize the expression.
- **Accumulation of errors in the parsing process:** Errors in the recognition of individual symbols or sub-expressions can accumulate and propagate throughout the parsing process, leading to incorrect recognition of the entire expression.

Zhang et al. [1] first proposed an end-to-end offline recognition method, referred to as the WAP model. The WAP model uses a full convolutional network (FCN) as the encoder and a gated recurrent unit (GRU) as the decoder and applies a time-space hybrid attention mechanism in the decoding process to convert mathematical expression images directly into latex command sequences. Later, Zhang et al. improved the WAP model [3] by using a tightly connected convolutional network on the encoder, while proposing a multi-scale attention model to solve the recognition problem of mathematical symbols, and the results were well improved. In addition to the encoder–decoder method above, there are some other methods. For example, in order to improve the robustness of the model, Wu et al. [4] proposed a method based on adversarial learning, which also achieved good results. The current mainstream models are built on an encoder–decoder framework, such as replacing the encoder with a transformer, which has recently become popular in natural language processing [5] or using a multi-headed attention mechanism in the decoder instead of the traditional spatial attention mechanism [6], or building models based on sequences of relations [7], all of which are based on improving the decoder or the form of the output sequence.

In our previous research, we proposed a model for offline handwritten mathematical expression recognition [8]. In this chapter, we further explore the performance improvement and extension of the model. Specifically, we have mainly carried out the following expansion work:

- **Multi-head attention experiment:** We introduced a multi-head attention mechanism to compare the impact of different types of attention modules on model performance.
- **Attention visualization:** We have also added an attention visualization process to help readers better understand how the model works during the decoding process. This visualization

demonstrates how the model assigns attention weights to different positions based on the content of the input image, thereby assisting the decoder in correctly parsing mathematical expressions.

Through these extensions, we aim to improve the performance of the model in offline handwritten mathematical expression recognition tasks and enrich our previous research content. In this chapter, we provide a detailed introduction to the experiments and results of these extended works and demonstrate how they can help improve the performance of our model.

8.1.2 Our main work

The encoder is the start of the encoder–decoder framework, and if it is poor, then it is difficult to get good results even if the decoder is nice. Therefore, we keep the decoder of the WAP model unchanged and investigate how to improve the encoder to make the extracted features more informative. First, we innovatively add residual attention modules at reasonable locations in the encoder, which greatly improves the performance of the encoder and hence the recognition accuracy. We also use the label smoothing method to change the original one-hot label into a soft label to assist the training, which enhances the generalization and calibration ability of the model. Finally, we conduct ablation experiments to compare the performance of different types and numbers of non-local attention modules in the handwritten expression recognition model. Our best model improved the accuracy by 6.32%/6.03%/7.44% over the original model (WAP) on the CROHME2014/2016/2019 test sets.

Next, we provide a comprehensive evaluation of the proposed method. We start with the good aspects of the method:

- **Introducing residual attention module:** Introducing the residual attention module in the encoder of the model enables the model to better capture global information. This improvement helps improve the performance of the model, especially when dealing with complex mathematical expressions.
- **Using label smoothing:** Using the label smoothing method as a loss function helps prevent overfitting of the model, improve generalization ability, and model calibration ability.
- **Conduct experiments on various types of residual attention modules and different levels:** By comparing residual

attention modules of different types and levels, find the most suitable configuration for the model.
- **Performance comparison with other HMER systems:** The superiority of the model has been demonstrated by comparing its performance with other offline handwritten mathematical expression recognition systems.

Furthermore, by adopting a self-critical perspective, we are aware that this chapter still has areas for improvement:

- **Exploration of attention modules with different levels:** Further research is needed on different types and quantities of attention modules to find the optimal configuration. More complex attention mechanisms can also be explored or adaptive attention methods can be introduced.
- **More complex decoder architecture:** It is possible to try using more complex decoder architectures, such as the transformer model, which is widely used in the field of natural language processing. Transformer has a self-attention mechanism that can handle long-distance dependencies and performs well in some sequence to sequence tasks. Introducing transformer or its variants into the decoder section may improve performance.
- **Integrating multi-scale information:** Mathematical expressions typically contain symbols and structures of different sizes and complexities. Decoders can better handle these situations by integrating multi-scale information. Consider introducing multi-scale attention mechanisms or using multi-level decoders.

8.2 Related Studies

There are currently two HMER methods: the grammar-based recognition method and the end-to-end deep learning method. In this section, we focus on the end-to-end deep learning method and related attention modules.

The mainstream deep learning framework is the encoder–decoder framework. The most typical model is the WAP model. In recent years, many models have been enhanced mainly by optimizing the decoder or the output sequence.

Improved WAP uses DenseNet as an encoder to extract features and two layers of unidirectional GRU network in the decoder to decode them. The HMER training samples are single channel images, and conventional network models mostly focus on three channel color images, but their performance is not good on single channel images. Since DenseNet has a dense connection structure that can better extract the features of single channel images, we chose DenseNet as our encoder, and the decoder in this chapter is the same as WAP.

8.2.1 *Attention*

Convolutional neural networks have certain limitations. The reason is that when we use convolution, the size of the convolution kernel is fixed, which leads to the lack of global information in the extracted features. Therefore, we add attention modules to the DenseNet as an encoder to achieve sufficient extraction of structural information from handwritten mathematical expressions. In this chapter, three types of attention modules are considered: multi-head attention [9], criss-cross attention [10], and coordinate attention [11]. By incorporating these attention modules into DenseNet encoder, we aim to achieve sufficient extraction of structural information from handwritten mathematical expressions. Our experiments show that the use of attention modules can improve the performance of our model on this task.

8.2.1.1 *Multi-head attention*

Multi-head attention is a powerful extension of the attention mechanism that has been successfully used in natural language processing tasks. In this approach, multiple sets of attention weights are learned to capture different semantic information from the input sequence. Each set of weights generates a context vector, and the final output is obtained by concatenating and linearly transforming these vectors. This allows the model to effectively extract information from the input sequence.

8.2.1.2 *Criss-cross attention*

For feature vectors, the general non-local attention module calculates the interrelationship between the target pixel and all other points in the feature map and uses this interrelationship to weight the features of the target pixel to obtain a more efficient target feature. In contrast, the criss-cross attention module only calculates the relationship between the target pixel and the pixel in the row and column in which it is located. A global contextual feature vector can be obtained after two identical criss-cross attention modules, which have a smaller number of parameters and are less computationally intensive than the usual non-local attention modules but yield more effective contextual information, maintaining long-range spatial dependence.

8.2.1.3 Coordinate attention

Coordinate attention is an improved version of channel attention Senet [12], in which global pooling is typically used to globally encode spatial information, but it compresses global spatial information into channel descriptors, making it difficult to retain positional information, which is crucial for capturing spatial structure in visual tasks. To encourage the attention module to capture remote interactions spatially with precise location information, coordinate attention decomposes global pooling into X-directional averaged pooling and Y-directional averaged pooling and finally fuses them to obtain global attention.

8.3 Methodology

This chapter is based on a multi-scale attention mechanism coding and decoding network for the recognition of offline handwritten mathematical expressions, starting with the encoding of the expressions using DenseNet and adding residual attention modules at appropriate locations.

8.3.1 *DenseNet encoder*

In the encoder section, DenseNet is used as a feature extractor for HME images, as DenseNet has a densely connected structure that allows for significant compression of the parameters that need to be trained, thus greatly reducing the complexity of model training and the time required. The subsequent experimental part of this chapter uses the DenseNet structure to encode expressions, producing better model results by adding residual attention modules to DenseNet.

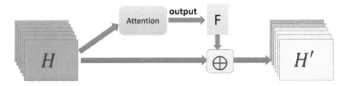

Fig. 8.1. Diagram of the residual attention network.

8.3.2 *Residual attention modules*

A limitation of CNN networks is the lack of global information in the extracted features. In this chapter, we refer to the residual attention network [13] to construct simple residual attention modules suitable for DenseNet (Figure 8.1).

For a feature map $H \in R^{C \times H \times W}$, an attention feature map $F \in R^{C \times H \times W}$ is obtained after the attention module.

$$F = \text{Attention}(H) \tag{8.1}$$

Similar to ResNet, the attention feature map F is directly added to the feature map to obtain a new feature map H':

$$H'(c, x, y) = H(c, x, y) + F(c, x, y). \tag{8.2}$$

The residual attention module enables the model to retain important features without destroying the properties of the original features, and the individual feature points of the resulting new feature map contain both local and global information to facilitate the decoder.

8.3.3 *DATWAP architecture*

8.3.3.1 *Encoder*

The design of the DAT network is relatively simple: The residual attention module is added after Transition (2) and Dense Block (3), respectively (Figure 8.2). In this chapter, the residual attention module is built using the multi-head attention module, criss-cross attention module, and the coordinate attention module, respectively.

Layers	Input size	Output size	DenseNet-B	DAT Net
Convolutions	$H \times W \times 1$	$\frac{H}{2} \times \frac{W}{2} \times 48$	7×7 conv, stride 2	
Pooling	$\frac{H}{2} \times \frac{W}{2} \times 48$	$\frac{H}{4} \times \frac{W}{4} \times 48$	Max pool, stride 2	
Dense block (1)	$\frac{H}{4} \times \frac{W}{4} \times 48$	$\frac{H}{4} \times \frac{W}{4} \times 432$	$\begin{bmatrix} 1 \times 1\ conv \\ 3 \times 3\ conv \end{bmatrix} \times 16$	
Transition (1)	$\frac{H}{4} \times \frac{W}{4} \times 432$	$\frac{H}{4} \times \frac{W}{4} \times 216$	1×1 conv	
	$\frac{H}{4} \times \frac{W}{4} \times 216$	$\frac{H}{8} \times \frac{W}{8} \times 216$	2×2 average pool, stride 2	
Dense block (2)	$\frac{H}{8} \times \frac{W}{8} \times 216$	$\frac{H}{8} \times \frac{W}{8} \times 600$	$\begin{bmatrix} 1 \times 1\ conv \\ 3 \times 3\ conv \end{bmatrix} \times 16$	
Transition (2)	$\frac{H}{8} \times \frac{W}{8} \times 600$	$\frac{H}{8} \times \frac{W}{8} \times 300$	1×1 conv	
	$\frac{H}{8} \times \frac{W}{8} \times 300$	$\frac{H}{16} \times \frac{W}{16} \times 300$	2×2 average pool, stride 2	
Attention (1)	$\frac{H}{16} \times \frac{W}{16} \times 300$	$\frac{H}{16} \times \frac{W}{16} \times 300$	--	Residual attention
Dense block (3)	$\frac{H}{16} \times \frac{W}{16} \times 300$	$\frac{H}{16} \times \frac{W}{16} \times 684$	$\begin{bmatrix} 1 \times 1\ conv \\ 3 \times 3\ conv \end{bmatrix} \times 16$	
Attention (2)	$\frac{H}{16} \times \frac{W}{16} \times 684$	$\frac{H}{16} \times \frac{W}{16} \times 684$	--	Residual attention

Fig. 8.2. DenseNet architectures for WAP and DenseNet architectures with attention (DAT) for our model.

8.3.3.2 Decoder

Given an input mathematical expression image, let $X \in R^{D \times H \times W}$ denote extracted DATNet feature maps, where D, H, and W represent channel number, height, and width, respectively. We consider these feature maps as a sequence of feature vectors $x = x_1, x_2, \ldots, x_L$, where $x_i \in R^D$ and $L = H \times W$. Let $y = y_1, y_2, \ldots, y_T$ denote the target output sequence, where y_T represents a special end-of-sentence (EOS) symbol. The decoder uses two unidirectional GRU layers and an attention with cover vector mechanism. The specific calculation formula is as follows:

$$h_t^{(0)} = GRU^{(0)}(y'_{t-1}, h_{t-1}^{(1)}), \qquad (8.3)$$

$$c_t = f_{\text{att}}(h_t^{(0)}, x), \qquad (8.4)$$

$$h_t^{(1)} = GRU^{(1)}(c_t, h_t^{(0)}). \qquad (8.5)$$

Previous step's y'_{t-1}, which is a trainable embedding of y_{t-1}, and output $h_{t-1}^{(1)}$ are used as inputs in the first GRU layer to produce $h_t^{(0)} \in R^{d_{\text{GRU}}}$, which works as the query of the attention mechanism. Then, the attention with cover vector mechanism uses x as key and value to produce context $c_t \in R^{d_{\text{out}}}$. Finally, the second GRU layer uses c_t and $h_t^{(0)}$ as input and produces $h_t^{(1)} \in R^{d_{\text{GRU}}}$.

8.3.4 *Label smoothing*

Label smoothing combines uniform distribution, replacing the traditional one-hot label vector with a soft label vector y^{SL}:

$$y_i^{\text{SL}} = \begin{cases} 1-\alpha, & i = \text{target}, \\ \dfrac{\alpha}{K-1}, & i \neq \text{target}, \end{cases}$$

where K is the number of latex characters predicted and α is a small hyperparameter. Deep neural networks tend to overfit during training, which can reduce their generalization ability. In addition, large datasets often contain mislabeled data, meaning that neural networks should inherently be skeptical of the 'right answer'. Using label smoothing can produce a better-calibrated network, which will generalize better and ultimately produce more accurate predictions on the test data. In particular, in seq2seq tasks where beam search is taken for decoding, training with smooth loss gives a good boost. During the training process, our encoder and decoder both have methods such as early stop, drop out, and batch normalization to prevent overfitting and accelerate model convergence, as these methods are commonly used and are not discussed in detail here.

8.4 Experiments

8.4.1 *Experimental setup*

We evaluated our method on the CROHME dataset [14–16]. The training set chosen for this chapter is the CROHME2014 training set, containing a total of 8836 handwritten InkML files of mathematical expressions, which we converted into the corresponding 8836 offline images as the training set. The CROHME2014 test set with 986 images was chosen as the validation set. The CROHME2016 test set containing 1147 images and the CROHME2019 test set containing 1199 images were used as the test sets. The experiments were run on an Ubuntu server with Pytorch version 1.7.1 and Python version 3.7.9. The loss function used in the experiments was smooth loss, label smoothing adjustment $\epsilon = 0.1$. The model is trained using the AdaDelta optimizer with an initial learning rate of 1 and weight decay of 0.0001, patience = 15, and the learning rate is reduced when the error rate does not decrease after 15 epochs. After the learning rate has been reduced twice, the training is terminated early to prevent overfitting if the results do not improve on the validation set. Gradient clipping is performed during training to prevent gradient explosion, and the clipping value is set to 100. Beam search is used for prediction, and the beam width is set to 10.

8.4.2 *Ablation experiment*

8.4.2.1 *Evaluation metrics*

ExpRate represents that the recognition of mathematical expressions in the image is completely correct, that is, the recognition result output by the model and the editing distance of the label are 0. In addition, "≤1 error" and "≤2 error" represent that the editing distance of the previous two is less than or equal to 1 and less than or equal to 2, respectively.

8.4.2.2 *Different types of residual attention modules*

We first evaluated the performance of the DATWAP model based on multi-head attention (see Table 8.1), criss-cross attention, and coordinate attention:

- DATWAP model based on criss-cross attention: attention (1) and attention (2) are criss-cross residual attentions.
- DATWAP model based on coordinate attention: attention (1) and attention (2) are coordinate residual attention.
- DATWAP model based on multi-head attention: attention (1) and attention (2) are multi-head attention.

Table 8.1. Performance comparison of DATWAP with criss-cross attention and DATWAP with coordinate attention on CROHME 2014, CROHME 2016, and CROHME 2019.

Model	2014 ExpRate (%)	2016 ExpRate (%)	2019 ExpRate (%)
WAP (Our implementation)	48.4	46.8	45.1
DATWAP with criss-cross	54.72	52.83	52.54
DATWAP with coordinate	54.11	49.26	48.62
DATWAP with multi-head	53.44	51.26	52.12

8.4.2.3 *Different layers of residual attention modules*

We try to add different layers of residual attention modules to DenseNet-B:

- **Add two layers of residual attention modules:** Residual attention module is added after Transition (2) and Dense Block (3) in Figure 8.3, respectively.
- **Add three layers of residual attention modules:** Residual attention module is added after Transition (1), Transition (2), and Dense Block (3) in Figure 8.3, respectively.

As can be seen from Table 8.2, the use of two layers of residual attention modules works best, which is another reflection of the fact that more attention modules are not better and that the location and number of layers of attention modules used are important.

8.4.3 *Comparison with other HMER systems and case study*

In Tables 8.3 and 8.4, we compare our models with other offline HMER systems on the CROHME2016 and CROHME2019 test sets, respectively. Performance comparison of a single model on the CROHME2016/2019 test set (%).

As can be seen from Tables 8.3 and 8.4, adding two layers of residual attention modules at the appropriate locations in Dense-B while

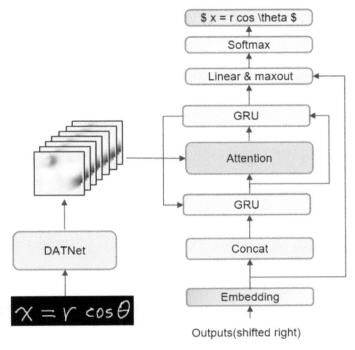

Fig. 8.3. Diagram of the overall architecture of DATWAP.

Table 8.2. Performance comparison of residual attention modules in different layers of the DATWAP with on CROHME 2014, CROHME 2016, and CROHME 2019.

Residual attention module layers	2014 ExpRate (%)	2016 ExpRate (%)	2019 ExpRate (%)
2	54.72	52.83	52.54
3	53.4	50.39	50.04

training the model using smooth loss, does result in good recognition. The experiments demonstrate that our model (DATWAP) improves the recognition rate of WAP expressions from 48.4%/46.8%/45.1% to 54.72%/52.83%/52.54% on the CROHME2014/2016/2019 test sets.

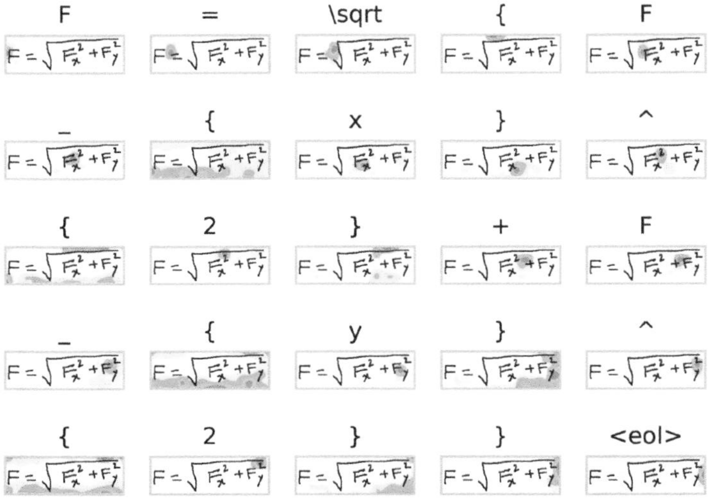

Fig. 8.4. Visualization of the attention process.

Table 8.3. Performance comparison of HMER systems on CROHME 2016.

Model	ExpRate (%)	≤1 error (%)	≤2 error (%)
Wiris [15]	49.61	60.42	64.69
PGS [17]	45.60	62.25	70.44
WAP [18]	46.82	64.64	65.48
WS-WAP [19]	48.91	57.63	60.33
PAL-v2 [4]	49.61	64.08	70.27
BTTR [5]	52.31	63.90	68.61
Relation-based Rec-wsl [7]	52.14	63.21	69.40
DATWAP(our)	52.83	65.65	70.96

8.4.4 Visualization and case analysis

Figure 8.4 illustrates the attention-based decoding process, where mintcream is the background, black is the font, and other color areas are the focus areas of attention. The darker the color, the higher the attention weight. It can be seen that the attention can not only capture the spatial position of each character but also use the spatial structure information to assist the decoder in parsing symbols '' and ''.

Table 8.4. Performance comparison of HMER systems on CROHME 2019.

Model	ExpRate (%)	≤1 error (%)	≤2 error (%)
Univ. Linz [16]	41.29	54.13	58.88
DenseWAP [20]	41.7	55.5	59.3
DenseWAP-TD [20]	51.4	66.1	69.1
BTTR [5]	52.96	65.97	69.14
Relation-based Rec-wsl [7]	53.12	63.89	68.47
DATWAP(our)	52.54	65.72	70.81

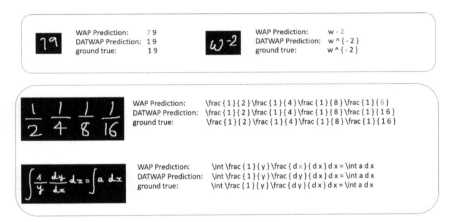

Fig. 8.5. Case studies for the "WAP" [1] and "DATWAP" (Ours). The red symbols represent incorrect predictions.

As can be seen from Figure 8.5, the accuracy is greatly improved relative to WAP as the encoder used by DATWAP extracts features with richer global information, facilitating the decoder.

8.5 Conclusion and Outlook for the Future

This chapter is based on an improvement of the WAP model. Using the decoder of WAP, the model performs better in recognizing handwritten mathematical expressions by introducing a residual attention module into the encoder. Since the CNN focuses on local information, while the residual attention module focuses on global information, the addition of the residual attention module can make full use of the long-distance correlation and alleviate the problem that the dependency information between the captured current symbol and the previous symbol diminishes with increasing distance. The model is also trained using smoothing losses, which is similar to adding noise to the original label, effectively preventing overfitting. Experiments conducted on the CROHME dataset show that our model (DATWAP) improves the recognition rate of WAP expressions from 48.4%/46.8%/45.1% to 54.72%/52.83%/52.54% on the CROHME 2014/2016/2019 test sets.

As mentioned in the previous model summary, there is still a lot of room for improvement in HMER in the future:

- **For decoder:** It is possible to consider replacing the decoder with a transformer model decoder.
- To address the issue of insufficient training data, contrastive learning can be considered to pre-train the encoder or using AIGC related models to generate images containing mathematical expressions.
- For multi-scale mathematical expressions, it is possible to consider introducing multi-scale attention mechanisms or using multi-level decoders.

As researchers of HMER, we sincerely hope that HMER-related methods can be continuously improved and play a greater role in real life.

References

[1] J. Zhang, J. Du, S. Zhang, D. Liu, Y. Hu, J. Hu, S. Wei and L. Dai, Watch, attend and parse: An end-to-end neural network based approach to handwritten mathematical expression recognition, *Pattern Recognition* **71**, 196–206 (2017).

[2] K. F. Chan and D. Y. Yeung, Mathematical expression recognition: A survey, *International Journal on Document Analysis and Recognition* **3**, 3–15 (2000).

[3] J. Zhang, J. Du and L. Dai, Multi-scale attention with dense encoder for handwritten mathematical expression recognition, in

2018 24th international conference on pattern recognition (ICPR). IEEE, pp. 2245–2250 (2018).

[4] J. W. Wu, F. Yin, Y. M. Zhang, X. Y. Zhang and C. L. Liu, Handwritten mathematical expression recognition via paired adversarial learning, *International Journal of Computer Vision* **128**(10), 2386–2401 (2020).

[5] W. Zhao, L. Gao, Z. Yan, S. Peng, L. Du and Z. Zhang, Handwritten mathematical expression recognition with bidirectionally trained transformer, in *2021 International Conference on Document Analysis and Recognition (ICDAR)*. Springer, pp. 570–584 (2021).

[6] H. Ding, K. Chen and Q. Huo, An encoder-decoder approach to handwritten mathematical expression recognition with multi-head attention and stacked decoder, in *2021 International Conference on Document Analysis and Recognition (ICDAR)*. Springer, pp. 602–616 (2021).

[7] T. N. Truong, H. Q. Ung, H. T. Nguyen, C. T. Nguyen and M. Nakagawa, Relation-based representation for handwritten mathematical expression recognition, in *2021 International Conference on Document Analysis and Recognition (ICDAR)*. Springer, pp. 7–19 (2021).

[8] Q. Lin, C. Wang, N. Bi, C. Y. Suen and J. Tan, An encoder-decoder approach to offline handwritten mathematical expression recognition with residual attention, in *Pattern Recognition and Artificial Intelligence - 3rd International Conference, ICPRAI 2022*, Paris, France, June 1–3, 2022, Proceedings, Part I. Lecture Notes in Computer Science, Vol. 13363. Springer, pp. 335–345 (2022).

[9] A. Vaswani, N. Shazeer, N. Parmar, J. Uszkoreit, L. Jones, A. N. Gomez, Ł. Kaiser and I. Polosukhin, Attention is all you need, *Advances in Neural Information Processing Systems*, Vol. 30 (2017).

[10] Z. Huang, X. Wang, L. Huang, C. Huang, Y. Wei and W. Liu, Ccnet: Criss-cross attention for semantic segmentation, in *Proceedings of the IEEE/CVF International Conference on Computer Vision*, pp. 603–612 (2019).

[11] Q. Hou, D. Zhou and J. Feng, Coordinate attention for efficient mobile network design, in *Proceedings of the IEEE/CVF Conference on Computer Vision and Pattern Recognition*, pp. 13713–13722 (2021).

[12] J. Hu, L. Shen and G. Sun, Squeeze-and-excitation networks, in *Proceedings of the IEEE Conference on Computer Vision and Pattern Recognition*, pp. 7132–7141 (2018).

[13] F. Wang, M. Jiang, C. Qian, S. Yang, C. Li, H. Zhang, X. Wang and X. Tang, Residual attention network for image classification, in

- [14] H. Mouchere, C. Viard Gaudin, R. Zanibbi and U. Garain, ICFHR 2014 competition on recognition of on-line handwritten mathematical expressions (CROHME 2014), in *2014 14th International Conference on Frontiers in Handwriting Recognition (ICFHR)*. IEEE, pp. 791–796 (2014).
- [15] H. Mouchère, C. Viard Gaudin, R. Zanibbi and U. Garain, ICFHR2016 CROHME: Competition on recognition of online handwritten mathematical expressions, in *2016 15th International Conference on Frontiers in Handwriting Recognition (ICFHR)*. IEEE, pp. 607–612 (2016).
- [16] M. Mahdavi, R. Zanibbi, H. Mouchere, C. Viard Gaudin and U. Garain, ICDAR 2019 CROHME+ TFD: Competition on recognition of handwritten mathematical expressions and typeset formula detection, in *2019 International Conference on Document Analysis and Recognition (ICDAR)*. IEEE, pp. 1533–1538 (2019).
- [17] A. D. Le, B. Indurkhya and M. Nakagawa, Pattern generation strategies for improving recognition of handwritten mathematical expressions, *Pattern Recognition Letters* **128**, 255–262 (2019).
- [18] J. Wang, J. Du, J. Zhang and Z.-R. Wang, Multi-modal attention network for handwritten mathematical expression recognition, in *2019 International Conference on Document Analysis and Recognition (ICDAR)*. IEEE, pp. 1181–1186 (2019).
- [19] T. N. Truong, C. T. Nguyen, K. M. Phan and M. Nakagawa, Improvement of end-to-end offline handwritten mathematical expression recognition by weakly supervised learning, in *2020 17th International Conference on Frontiers in Handwriting Recognition (ICFHR)*. IEEE, pp. 181–186 (2020).
- [20] J. Zhang, J. Du, Y. Yang, Y.-Z. Song, S. Wei and L. Dai, A tree-structured decoder for image-to-markup generation, in *International Conference on Machine Learning*. PMLR, pp. 11076–11085 (2020).

Chapter 9

A Complexity Analysis on the General Feasibility of Patch-Based Adversarial Attacks on Semantic Segmentation Problems

András Horváth and Soma Kontár

*Faculty of Information Technology and Bionics,
Peter Pazmany Catholic University,
Práter u. 50/A, Budapest, Hungary*

Abstract

Deep neural networks have found successful application across a wide range of tasks. However, in safety-critical scenarios, the persistent threat of adversarial attacks remains a concern. These attacks have been demonstrated across different tasks involving classification and detection. They are typically characterized by their capacity to generate outputs that can significantly disrupt network functionality.

This research endeavors to showcase the efficacy of patch-based attacks in modifying the outcomes of segmentation networks. This study encompasses both simulated scenarios and real-world instances to illustrate this phenomenon. By exploring several illustrative cases and delving into network intricacies, we reveal that the scope of feasible output maps achievable through patch-based attacks of a given size tends to be more constrained than the corresponding affected area. This discrepancy holds true especially in practical applications, where certain areas are targeted for attack.

Through our analysis, we establish that the practical applicability of most patch-based attacks is limited. These attacks lack the ability to universally generate arbitrary output maps. Even in cases where such generation is feasible, the spatial extent of these attacks is confined and notably smaller than the patches' receptive field.

9.1 Introduction

As the integration of deep neural networks into our daily routines gains widespread traction, concerns surrounding the resilience and dependability of these networks grow increasingly prominent. The exploration of adversarial attacks, which exploit the vulnerabilities inherent in neural networks, has garnered substantial attention in recent years. These attacks possess a general nature, as adept optimization methods empower them to produce arbitrary outputs, irrespective of the input image. Consequently, their potential impact in practical applications looms large.

The inception of adversarial attacks can be traced back to their introduction in [1]. These attacks unearth a pivotal aspect of deep neural networks: While these networks exhibit adept generalization not only across customary input datasets but also among akin inputs, they remain susceptible to manipulation by malicious agents. This vulnerability stems from the high dimensionality of inputs, enabling the creation of non-realistic samples that deviate drastically from both human judgment and expected outcomes.

Subsequent years witnessed a thorough exploration of the exploitative capabilities of adversarial attacks, building upon the foundational work of the original authors [2–5]. Novel attack strategies emerged, refining the potency of generated attacks [6, 7]. Advancements included the development of black-box attacks, where gradients of the network are dispensable [8–10].

Later investigations expanded the purview of adversarial attacks beyond mere classification, encompassing more intricate tasks like detection and localization [11]. These attacks were also demonstrated across diverse network architectures, such as Faster-RCNN [12].

Initially, the earliest attacks tackling classification and detection challenges involved applying subtle, low-intensity distortions across the entirety of images, as exemplified in [2] for classification tasks. However, subsequent investigations revealed the frailty of such low-intensity attacks when confronted with real-world scenarios [13]. In practical settings, this specialized additive noise is often subject to modifications from perspective distortions and environmental factors, such as additive noise (e.g., changes in illumination). Moreover, it lacks authenticity since attackers typically require access to image processing systems to manipulate all components of the input rather than altering real-world objects themselves.

While the realm of classification and detection witnessed the emergence of authentic, robust attacks, the domain of segmentation problems predominantly saw scrutiny directed at low-intensity attacks [14, 15]. Segmentation tasks exhibit greater complexity, as their outcomes hinge on intricate aspects of input samples. In classification, the anticipated outcome should remain unaffected by an object's pose, and in detection, only marginal alterations are expected. However, with segmentation, even slight rotations or pose variations (e.g., a person moving their arm) can potentially lead to substantial changes in segmentation masks. This suggests that segmentation networks might display heightened resilience against real-world adversarial assaults.

Research such as [14, 15] demonstrated that networks trained for semantic segmentation tasks can indeed fall prey to low-intensity noise attacks, enabling authors to produce diverse output maps through controlled additive noise. While these methods haven't definitively established their capacity to generate arbitrary output maps, they have shown the potential to yield highly uncorrelated and randomly selected output maps. This outcome has fostered a prevailing consensus within the scientific community that these low-intensity strategies hold the potential to yield arbitrary output maps.

In the study conducted by Nakka *et al.* [16], the vulnerability of state-of-the-art semantic segmentation networks to certain indirect local adversarial attacks is showcased. This attack strategy involves placing a patch within the environment, consequently creating "dead zones" for a specific class of objects. While this underscores the susceptibility of some networks to patch-based adversarial attacks, the authors observed that models with larger fields of view are more responsive to such attacks. In contrast, our approach more closely mirrors real-life situations, as we exclusively modify the targeted object without requiring access to the encompassing environment in which the attack is executed.

Remarkably, the authors of this chapter are not aware of any successful patch-based attacks on semantic segmentation problems. This absence underscores the intricacy of generating such samples.

Within this chapter, we intend to establish the feasibility of employing patch-based attacks within semantic segmentation problems. Subsequent to demonstrating their feasibility, we delve into an analysis of their generality. Frequently, the repertoire of attainable output maps, employing patches of limited dimensions, tends to be rather confined. Consequently, these attacks, contrary to the prevailing assumption, lack the capacity to generate arbitrary outputs. This realization drives the conclusion that their generality is restricted.

Nevertheless, it is important to acknowledge that the lack of generality in patch-based attacks for segmentation challenges does not preclude their practical applicability. Rather, this denotes that they cannot universally generate arbitrary outputs. Nonetheless, the question of which specific outputs can be generated by adversarial patches and which cannot remains unresolved. This inquiry is slated for exploration in our forthcoming research endeavors.

9.2 Adversarial Attacks

The concept of an *adversarial example* was introduced in [1]. In this seminal work, attacks were devised for neural networks tailored to image classification. These attacks were crafted using exceedingly subtle, specially crafted additive noise, which remained entirely imperceptible to the human eye. The approach employed to generate these adversarial examples revolved around maximizing the network's response to a specific class by manipulating the input image.

The initial wave of attacks, as presented in [2], leveraged the gradient of the cost function (J) concerning both the input (x) and the anticipated output (y). This gradient's elements were multiplied by a constant factor, scaling the intensity of the noise (expressed formally as ϵ sign $\nabla_x J(\boldsymbol{\theta}, \boldsymbol{x}, y)$), where $\boldsymbol{\theta}$ symbolizes the model's parameter vector). This approach significantly expedited the attack generation process and came to be known as the Fast Gradient Sign Method (FGSM).

An expansion of the FGSM framework, introduced in [3], entailed employing not solely the sign of the raw gradient of the loss but also a scaled version of the gradient's raw magnitude. This technique is commonly referred to as the fast gradient value (FGV) method.

An additional enhancement to the iterative variant of FGSM, introduced in [4], entailed the incorporation of momentum into the equation. This augmentation was grounded in the idea that, analogous to conventional optimization during training, momentum could potentially assist in circumventing suboptimal local minima and other non-convex configurations in the objective function's landscape.

Building upon the premise that the resilience of a binary classifier f at a specific point x_0 equates to the distance between x_0 and the separating hyperplane $\Delta(x_0; f)$, [5] developed their approach. Accordingly, the smallest necessary perturbation to alter the output sign of f corresponds to the orthogonal projection of x_0 onto the separating hyperplane. They mathematically solve this challenge in a closed-form formulation and subsequently apply these minor perturbations iteratively to the image until the classifier's decision shifts. This approach was subsequently extended to accommodate multi-class classification problems as well.

Despite the significance of these methodologies from a theoretical standpoint and the generality of their generation techniques, they do not substantially jeopardize the practical applicability of neural networks. This is due to their imposition of constraints on the amount of introduced noise. The introduction of even the slightest perturbations, aside from the engineered additive noise, such as shifts in perspective, alterations in illumination, or lens distortions, could profoundly disrupt the desired outcomes. As a result, the practical deployment of these attacks in real-world scenarios remains implausible [13].

In [6,7], robust and real-world attacks targeting various classification networks were introduced. These methodologies involve the creation of adversarial patches. In contrast to the preceding approaches that relied on global yet low-intensity alterations, these methods induce distortions within localized regions, albeit without constraining intensity values.[a] Notably, successful adversarial patch attacks were even demonstrated using exclusively black and white patches [8],

---

[a]Except within the overall image value bounds.

where the optimization focuses on the locations and sizes of stickers rather than the patch intensities. These attacks, bypassing the necessity for network gradients during optimization, open avenues for black-box attacks [9, 10]. Here, attackers require access solely to final responses and confidence values to devise attacks via evolutionary algorithms.

Subsequently, these techniques extended their reach to encompass detection and localization challenges [11], incorporating various network architectures, such as Faster-RCNN [12].

For a comprehensive survey of adversarial attacks, offering more intricate descriptions of most of the aforementioned methods, one can refer to the overview provided in [17].

The resilience of segmentation networks against adversarial attacks has received extensive scrutiny in recent years [14, 15, 18, 19]. However, the focus has predominantly rested on global, low-intensity attacks. Remarkably, there appears to be no existing publication demonstrating the application of patch-based attacks on semantic segmentation tasks.

9.3 Patch-Based Segmentation Attacks on a Simple Dataset

In our quest for a straightforward segmentation dataset akin to MNIST [20] for classification, we encountered a scarcity of options. Consequently, we took the initiative to craft our own dataset, drawing inspiration from CLEVR [21]. Although the original CLEVR dataset was primarily tailored for visual question answering and lacked segmentation masks, we adapted the dataset generator script. Through this modification, we successfully generated masks for semantic, amodal, and instance segmentation.

This bespoke dataset comprises 25,200 color images, each sized at 320 × 240 pixels. These images feature uncomplicated objects, accompanied by corresponding instance masks, amodal masks, pairwise occlusions, and three-dimensional coordinates for each object. This compilation offers a versatile resource for tackling diverse tasks, encompassing three-dimensional reconstruction, instance segmentation, and amodal segmentation. The dataset showcases objects characterized by simple shapes while also incorporating elements like shadows, reflections, and diverse illuminations. This diversity renders the dataset pertinent for evaluating the efficacy of segmentation algorithms.

For the convenience of fellow researchers, we have made the dataset a generator script, and all associated training codes available in a dedicated repository.[b] This repository facilitates the replication of our experiments and enables a comprehensive exploration of the applied parameters.

We opted for the U-net architecture [22] to train on our custom CLEVR-inspired dataset. Our objective was to showcase adversarial attacks within architectures that manage classification and segmentation within the same layers rather than using distinct heads as seen in Mask-RCNN [23]. In cases like Mask-RCNN, the attack might deceive the classifier head while still maintaining a comparable instance mask. To maintain this aspect, we adopted a U-Net-like structure composed of convolution blocks with channel configurations of 8, 16, 32, and 64. We employed strided convolution for downsampling and transposed convolution for upsampling.

For training, we utilized 23,400 images while reserving 1,800 images for validation. Importantly, the validation scenes were independently generated from the training scenes. Our focus lay in semantic segmentation, where the network's task was to produce a four-channel output image representing the pixel probabilities for belonging to a cube, sphere, cylinder, or the background.

To evaluate adversarial attacks, we randomly selected 100 samples from our test set and employed the method outlined in [4] to train adversarial patches. We explored scripted attacks that involved altering the expected output class of a chosen object while maintaining the mask shape. However, even in this scenario, the network did not benefit from prior learned shapes or the consistent shapes in our database. This emphasized the challenge of segmenting spheres masquerading as cubes or cylinders.

To demonstrate the network's capacity for arbitrary shape segmentation, we manually crafted expected masks and tested our approach on them. For larger-scale experiments, we leaned toward

---

[b]https://drive.google.com/drive/folders/1UzozXfsuOb-IpUGF-9YrfIVN1Wrr AWsw.

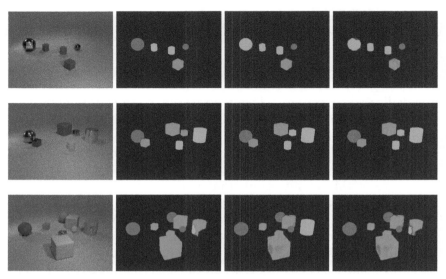

Fig. 9.1. This illustration presents segmentation attacks carried out on the CLEVER dataset. In the initial column, the images are showcased after the patch has been optimized specifically for these samples over 5000 iterations. Subsequently, the second column displays the anticipated outputs post-attack, while the third and fourth columns showcase the network's outputs before and after the attack, respectively. The top two rows feature samples where object classes were swapped, while the bottom row showcases an instance where the mask was manually crafted. It is important to observe that sections of the output mask have been removed to highlight that the modifications entail not only transforming objects into different classes but also potentially causing partial or complete disappearance. Additionally, we underline that there exists no theoretical distinction between altering an output pixel to align with a desired class or leaving it unassigned to any class.

the scripted method for class switching due to its easy implementation through scripts.

For a visual representation of sample attacks, the expected masks, and the network outputs before and after the attacks, refer to Figure 9.1.

As it can be seen from the previous examples, patch based attacks were possible on this simple segmentation task. The outcome of the network was close to the expected mask in almost all examined cases and even though they were not perfectly reconstructed every time, altogether 95% of the output pixels were modified as expected.

9.3.1 Real-life images of simple shapes

After successfully generating attacks in simulation, we proceeded to acquire 100 real-life samples from a variety of setups (10 distinct configurations from 10 different viewpoints) for testing our methodology. In this real-world testing, we deployed our method without undertaking any fine-tuning or additional optimization on the genuine samples. Furthermore, we refrained from implementing domain adaptation techniques [24, 25]. Remarkably, our network exhibited satisfactory performance on real-life samples. Despite encountering noise in the segmentation results and the emergence of minor objects in the background, the segmentation of authentic objects remained accurate in terms of both shapes and classes. These attributes hold utmost significance in scrutinizing potential attack scenarios.

To advance toward more intricate and robust validation, we ventured into the realm of real-life images featuring tangible yet straightforward objects like wooden blocks. Subsequently, we subjected these real images to attack simulations, where we optimized patches for them. In essence, this procedure involved capturing images without stickers, followed by the placement and optimization of stickers within simulations based on the acquired images. To offer a qualitative glimpse into how these patches could manipulate the network's output and induce alterations in the output classes of specific objects, we present illustrative samples in Figure 9.2.

The preceding experiments underscore the vulnerability of segmentation networks to patch-based attacks. These attacks can only be crafted when the attacker possesses access to the system's processing pipeline and direct control over pixel values. In contrast, it was established in [8] that patches can be physically and directly affixed to objects, proving their resilience and practical utility.

To further explore this, we hand-modified the outputs of 10 selected images. By repainting one of the objects on the segmented output map and employing the same methodology as used with simulated data, we managed to alter the network's output on these specific images.

In an effort to demonstrate the robustness of patches within this experiment, we followed the patch generation approach introduced in [6]. For training purposes, we generated new simulated data

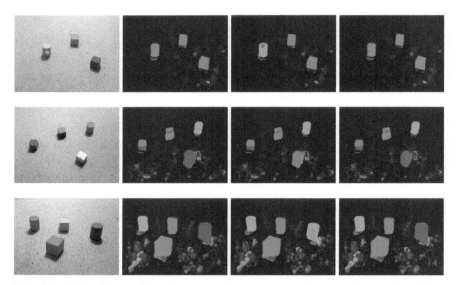

Fig. 9.2. This illustration showcases three segmentation attacks conducted on authentic images. In the leftmost column, the image is presented following the optimization of the patch for this specific sample over 5000 iterations. Subsequently, the middle column exhibits the anticipated outputs post-attack, while the third and fourth columns showcase the network's outputs before and after the attack, respectively.

that incorporated variances in view angles, scales (camera distances from objects), and lighting conditions. While the objects remained consistent across all images, these factors introduced variations. Subsequently, we endeavored to generate a single patch capable of effectively manipulating the chosen object, regardless of these aforementioned variances. In addition to this, we introduced minor random noise to the patch's intensity values and slightly adjusted the patch's position on the image to prevent the creation of low-intensity attacks. To further enhance the patch's consistency and effectiveness, we applied average pooling using a 3×3 kernel and stride one. This pooling approach enabled us to optimize the average of neighboring patch pixels within each kernel, offering more reliable and improved outcomes.

In all prior instances, patches were formulated and integrated into the image within simulation settings. However, in the present case, we adopted a distinct approach: We optimized a patch in simulation

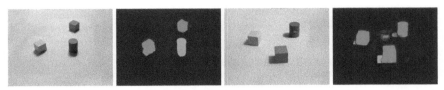

Fig. 9.3. This illustration demonstrates the impact of a physically printed patch in real-life segmentation scenarios. Notably, we successfully trained an adversarial patch through simulation, which then effectively transformed a cylinder into a cube within an authentic segmentation challenge.

to transform a cylinder into a cube within our simulated dataset and subsequently printed and evaluated it on real-life samples. An illustration of a randomly selected example from our sample set is displayed in Figure 9.3. As depicted, the blue cylinder was correctly segmented without the patch, but upon attaching the patch, its segmentation shifted to red, indicating its pixels were associated with a cube.

It's crucial to acknowledge that a more comprehensive investigation, involving a wider array of samples and more intricate scenarios, is necessary to comprehend the efficient generation of patch-based attacks in real-life contexts. However, our experiments effectively highlight the feasibility of robust adversarial patches. These patches possess the capability to remain effective across multiple viewpoints and conditions, showcasing their practical applicability.

9.4 Patch-Based Segmentation Attacks on Cityscapes

To delve into a more intricate case study, we turned our attention to the Cityscapes dataset [26], focusing on the dataset's semantic segmentation task. Within this context, we opted to work with the Deeplab V3 [27] and MobileNet V3 [28] architectures. A pivotal feature shared by both models is the integrated computation of mask prediction and classification, diverging from the approach employed by Mask-R-CNN [23].

Our experimental setup encompassed employing a ResNet-18 backbone for the Deeplab V3 architecture and the MobileNet V3 Large network with a segmentation head containing 128 filters.

Pre-trained models for both architectures were readily accessible: The Deeplab V3 model is accessible on Github, while the MobileNet V3 model can be obtained via PyPi.

To generate the adversarial patch, we adhered to the approach outlined in [6]. This method trains an adversarial patch within a white-box setting by minimizing the loss between a pre-defined target (often crafted manually) and the network's output. This is achieved by exclusively manipulating the values within the patch itself, which replaces a small portion of the image instead of perturbing every pixel with additive noise. Naturally, these values adhere to the standard image value bounds (e.g., [0, 255]) prior to preprocessing. In our specific experiments, we first selected a target class. For the sake of reproducibility, we employed an algorithm to determine the largest inscribable rectangle within a given binary object mask for the chosen class, a mask easily derived from the annotations in the Cityscapes dataset. We positioned the patch at the center of this targeted area, effectively confining the adversarial patch's impact to a quantifiable region. This approach aimed to mimic real-life scenarios where patches are typically positioned within object centers rather than spanning across multiple objects.

A selection of sample attacks is illustrated in Figure 9.4.

To assail the networks, we employed white-box patch-based attacks on the dataset's images, following a similar approach as with the toy dataset. The desired outputs were manually crafted, involving repainting a selected object on the actual output of the network. Similarly, the positions and sizes of the patches were also handcrafted, although no optimization was conducted regarding their specific dimensions and placements. For the patch value optimization process, we iterated over 10,000 steps.

Our attack generation targeted 50 randomly chosen images from the Cityscapes validation set for each network. Remarkably, the majority of these attacks proved successful. A collection of samples from these generated attacks can be observed in Figure 9.5. Intriguingly, we discovered that in most instances of successful attacks, the patch occupied less than 1% of the pixels within the targeted object. Examples illustrating this phenomenon alongside the corresponding patch sizes are displayed in Figures 9.6 and 9.7.

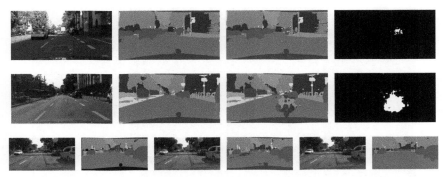

Fig. 9.4. Illustrative attacks on the Cityscapes dataset. The uppermost row provides insights into an attack on the DeeplabV3 architecture, wherein pixels associated with a car are transmuted into various other classes. The initial column portrays the input image featuring a 2 × 2 patch on the left lamp of a parked car, while the subsequent column showcases the original output of the trained network and the adversarial patch. The ensuing image reveals the network's output for the patched image along with the discernible effect of the patch. Pixels affected by the patch-induced change in output class (following argmax operation) are highlighted in white in the final image. Evidently, the 2 × 2 patch yields a notable impact, resulting in the alteration of output class for 7,461 pixels. The subsequent row offers a similar example but with the MobileNet V3 architecture. In this instance, an imaginary wall is superimposed in the middle of the road. The last row delves into an alternate attack scenario where a pedestrian is entirely erased from the network's output. The arrangement of this row, from left to right, is as follows: original input, target mask, Deeplab V3 patched input, Deeplab V3 attack result, MobileNet V3 patched input, and finally MobileNet V3 attack result. These examples not only underscore the susceptibility of neural networks employed in mission-critical domains, such as self-driving, but also emphasize the substantial threat posed by adversarial attacks in security-sensitive applications.

However, it's important to acknowledge that when dealing with larger objects, crafting patches of sufficiently small dimensions to remain inconspicuous while exerting a potent impact for a successful attack becomes challenging. An illustrative instance of this challenge is showcased in Figure 9.8.

As these results demonstrate, patch-based attacks are feasible in practice in case of segmentation problems. Based on this findings, one can ask the following question of what kind of limitations are

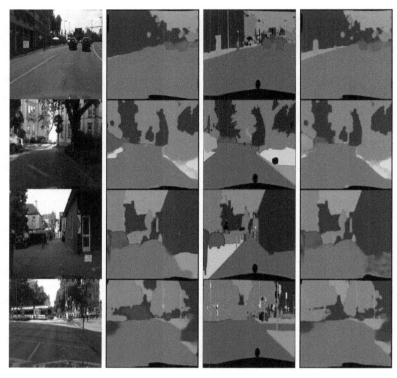

Fig. 9.5. This illustration presents the outcomes of the generated attacks on the Cityscapes dataset utilizing the U-Net inspired model, with each row representing an individual sample. The leftmost column showcases the images following the application of the attack patch. The subsequent column displays the network's response prior to the attack. In the third column, the (manually) modified mask employed during the attack is exhibited, while the final column reveals the network's output after the patch optimization. It's noteworthy that our experiments encompassed not only the switching of object classes but also the introduction of a non-existing object onto the sidewalk, as demonstrated in the topmost row.

there for patch based attacks: Can they generate arbitrary, general output maps?

9.5 A Complexity Analysis of Patch-Based Attacks

In this section, we establish the infeasibility of producing arbitrary output images through patch-based attacks on inputs. It's evident

A Complexity Analysis on the General Feasibility 219

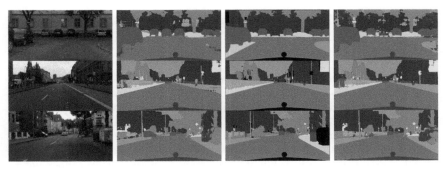

Fig. 9.6. In this figure, some further qualitative example attacks are depicted using the Deeplab V3 architecture.

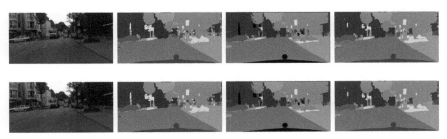

Fig. 9.7. This illustration offers a qualitative juxtaposition between the impact of an undersized and adequately sized patch. The arrangement of the figure is akin to previous illustrations: Each row corresponds to a single sample, while the columns portray the patched input image, the network's initial output before the attack, and the output following patch application, respectively. In the uppermost row, the scenario depicted highlights a case where the patch's dimensions are insufficient to yield a substantial effect.

that patch-based attacks are inherently limited by their inability to alter pixels falling beyond the receptive field of corresponding neurons, which is contingent on the chosen architecture. Here, we illustrate that if patch-based attacks were capable of yielding arbitrary outputs, the size of these patches would necessitate confinement to an area much smaller than their receptive field.

Our approach to proof is structured as follows: We undertake an examination of the myriad segmentation maps that can potentially emerge within a region, establishing an upper bound on their count based on the architecture of the network. Distinctness between two maps is determined by a discrepancy in the winning class — i.e., the

Fig. 9.8. This figure showcases the impact of a 30×30 patch on a notably sizable object. The initial image represents the unaltered input, followed by the network's output, the intended output post-attack, and finally, the network's response after the attack. Evidently, the patch size, in this case, is substantial enough to modify a considerable proportion of the targeted object's pixels. However, despite our meticulous efforts, this large patch size results in the object assuming various output classes post-attack. While it's important to note that we cannot definitively conclude that attacking semantic segmentation in this scenario (given the object-to-patch ratio) is entirely infeasible, it is apparent that methodologies with a certain complexity, successfully applied in classification and object detection tasks, do not yield similar success in this context.

class with the highest value post-softmax normalization — within at least one pixel. Such differences can only manifest through nonlinear alterations in the network's forward path. Consider an output map within a specific image region encompassing $W \times H$ pixels, where W denotes width and H signifies height of the region. Assuming a semantic segmentation network capable of discerning between D distinct classes, the potential output maps that can be generated amount to D^{WH} variations, wherein at least one pixel's classification differs.

For an attack to be capable of generating all these patterns, the network must encompass at least this quantity of distinct linear regions. Such distinctness is established through nonlinear alterations in network output, as changing the largest value at a designated output pixel necessitates non-monotonic shifts facilitated by nonlinear functions. In simple terms, the network needs to transition the output to different linear regions to achieve dissimilar output maps.

Furthermore, it's worth noting that not every linear region is mandated to generate a unique output map. While some of these regions could yield the same outcome, our forthcoming demonstration underscores that even if every individual region produced distinct outcomes, it still remains impractical to encompass all possible output maps.

To substantiate this, we compute the maximum count of linear regions engaged by the patch and subsequently reveal that this count significantly surpasses D^{WH}. This assertion underscores that, in practical terms, it is unattainable to generate arbitrary output maps in the context of semantic segmentation through patch-based attacks.

9.5.1 Upper bound of linear regions

The upper limit on the count of linear regions within a fully connected layer, incorporating ReLUs as nonlinearities, was initially introduced by Montufar et al. [29]. This theoretical framework posits an upper bound for the number of linear regions R_n as follows:

$$R_{L_{\text{FC}}} \leq \sum_{i=0}^{n_0} \binom{n_1}{i}, \tag{9.1}$$

where n_0 signifies the quantity of input neurons and n_1 denotes the number of output neurons within the layer.

In a subsequent paper in [30], this theorem was extended to convolutional networks. The authors demonstrated that the application of L consecutive convolutional layers cannot generate a greater number of distinct linear regions than

$$R_{\text{Conv}} \leq \prod_{l=1}^{L} \sum_{i=0}^{w_0 h_0 c_0} \binom{w_l h_l c_l}{i}. \tag{9.2}$$

In an L-layered network, where w_k and h_k denote the spatial dimensions of data in the k-th layer and c_k represents the number of channels in that layer, the formula correlates to the dimensions of the input data (e.g., w_0, h_0, and c_0 denote input data dimensions).

This equation implies that a convolutional layer operating on a 25×25 pixel input with 64 channels can potentially augment the number of linear regions by 3.29×10^{220}, while for 128 channels, this value increases to 5.3×10^{269}.

For patch-based attacks, the spatial dimensions are determined by the receptive field of the neurons encompassing the original patch. However, these dimensions tend to be relatively small to avoid detectability by human perception. In the context of a conventional convolutional layer with around 25×25 pixel patches and 128 or 256 channels, the multiplication of linear regions is capped at less than 10^{300}. As a result, a series of five convolutional layers could, at most, generate around 10^{1500} linear regions.

Considering 10 possible output classes, this implies that if the network were able to generate all potential output elements within a given region, the region's size could not exceed 10^{WH}, with WH having to be smaller than 1500. If not, the number of possible output maps would surpass the network's complexity.

As an example, a 25×25 patch would be constrained to generating output maps with an area of approximately 1500 pixels (roughly 38×38 pixels) if aiming for generality to yield arbitrary outputs.

Table 9.1 contains the maximal number of linear regions for networks which are typically applied on semantic segmentation problems. These data were calculated for different patch sizes along with the maximal patch size which can be generated with this complexity if we assume that patch generation is universal, namely arbitrary output maps can be created with an appropriate attack. The numbers were done considering the Cityscapes dataset as a case study, where each output pixel can belong to 19 different classes.

It is important to acknowledge that the calculations provided above represent upper bounds for the count of linear regions. The actual number of linear regions in practice is typically much smaller than the theoretical upper limit. Studies such as [32, 33] have shown that the number of linear regions can be measured or approximated

Table 9.1. This table displays the maximal number of nonlinear regions (R_N) for different network architectures (UNET [22], FCN8 [31], MobileNetv3-Large (MN_{V3}) [28], and DeeplabV3 with ResNet18 backbone (DL_{V3}) [27]) and patch sizes (S_R) along with maximal number of pixels which could be changed by such a region if generic output maps can be created.

	UNET	**FCN8**	MN_{V3}	DL_{V3}
R_N (2×2)	10^{219}	10^{168}	10^{229}	10^{584}
S_R (2×2)	13×13	11×11	13×13	21×21
R_N (5×5)	10^{1448}	10^{1203}	10^{1239}	10^{3421}
S_R (5×5)	33×33	30×30	31×31	51×51
R_N (10×10)	10^{5034}	10^{4646}	10^{3446}	10^{12725}
S_R (10×10)	62×62	60×60	51×51	99×99
R_N (20×20)	10^{16842}	10^{17864}	10^{9343}	10^{48151}
S_R (20×20)	114×114	118×118	85×85	194×194

through sampling in trained networks. In practical scenarios, the identified linear regions are notably fewer than the theoretical upper bounds. For instance, in the case of investigated seven and eight-layered convolutional networks, the upper bounds were 356180 and 819115 linear regions, respectively, yet through sampling methods, only 3398 and 4822 linear regions were detected in the trained networks.

This discrepancy between theoretical upper bounds and practical measurements suggests that the number of output maps that can be generated by a network is considerably smaller in real-world scenarios. Consequently, the maximum sizes in which arbitrary output maps can be produced are significantly reduced. Notably, the precise quantification of linear regions holds importance in practical applications for trained networks. However, the presented upper bounds are more general in nature and are solely dependent on the network architecture, unaffected by training data or network weights.

From the analyses conducted in the preceding sections, it's evident that adversarial patches have a more extensive impact on regions than what is indicated in Table 9.1. Based on these empirical findings, it can be concluded that practical patch-based attacks lack generality and are incapable of generating arbitrary output maps. It's crucial to clarify that this chapter's focus is on demonstrating the impossibility of generating arbitrary output maps. This conclusion doesn't imply that entire objects cannot be altered, non-existent objects cannot be generated, or individuals cannot be made to disappear in semantic segmentation using patches, as exemplified by some of our provided samples.

9.6 Conclusion

This chapter delved into the scope of patch-based attacks within the realm of semantic segmentation problems. Through our exploration, we examined a simplified synthetic dataset using the U-Net architecture, alongside the widely studied Cityscapes dataset, employing the DeepLab V3 and MobileNet V3 network architectures. Our investigation underscores the practical viability of patch-based attacks in semantic segmentation. Notably, even with diminutive patches like 2×2, these attacks can wield substantial influence, causing alterations to the output classes of segmentation maps over extensive areas, sometimes modifying more than 5000 pixels.

However, our in-depth complexity analysis of these architectures unveiled a crucial limitation. These segmentation networks possess a constrained capacity to generate distinct output maps. This revelation leads to the deduction that the efficacy of patch-based attacks in semantic segmentation hinges on specific conditions, largely contingent on the architecture of the network. Consequently, patch-based attacks do not hold the capability to yield arbitrary output maps. It's important to acknowledge that certain output maps in these segmentation networks remain immune to modification through patches of limited size, given a particular input configuration.

Acknowledgment

This research has been partially supported by the Hungarian Government by the following grant: 2018-1.2.1-NKP00008: Exploring the Mathematical Foundations of Artificial Intelligence and the support of the Alfréd Rényi Institute of Mathematics is also gratefully acknowledged.

References

[1] C. Szegedy, W. Zaremba, I. Sutskever, J. Bruna, D. Erhan, I. Goodfellow and R. Fergus, Intriguing properties of neural networks (2013), *arXiv preprint* arXiv:1312.6199.

[2] I. J. Goodfellow, J. Shlens and C. Szegedy, Explaining and harnessing adversarial examples (2014), *arXiv preprint* arXiv:1412.6572.

[3] A. Rozsa, E. M. Rudd and T. E. Boult, Adversarial diversity and hard positive generation, in *Proceedings of the IEEE Conference on Computer Vision and Pattern Recognition Workshops*, pp. 25–32 (2016).

[4] Y. Dong, F. Liao, T. Pang, H. Su, J. Zhu, X. Hu and J. Li, Boosting adversarial attacks with momentum, in *Proceedings of the IEEE Conference on Computer Vision and Pattern Recognition*, pp. 9185–9193 (2018).

[5] S.-M. Moosavi-Dezfooli, A. Fawzi and P. Frossard, Deepfool: A simple and accurate method to fool deep neural networks, in *Proceedings of the IEEE Conference on Computer Vision and Pattern Recognition*, pp. 2574–2582 (2016).

[6] T. B. Brown, D. Mané, A. Roy, M. Abadi and J. Gilmer, Adversarial patch (2017), *arXiv preprint* arXiv:1712.09665.

[7] A. Athalye, L. Engstrom, A. Ilyas and K. Kwok, Synthesizing robust adversarial examples (2017), *arXiv preprint* arXiv:1707.07397.

[8] K. Eykholt, I. Evtimov, E. Fernandes, B. Li, A. Rahmati, C. Xiao, A. Prakash, T. Kohno and D. Song, Robust physical-world attacks on deep learning models (2017), *arXiv preprint* arXiv:1707.08945.

[9] M. Alzantot, Y. Sharma, S. Chakraborty and M. Srivastava, Genattack: Practical black-box attacks with gradient-free optimization (2018), *arXiv preprint* arXiv:1805.11090.

[10] N. Papernot, P. McDaniel, I. Goodfellow, S. Jha, Z. B. Celik and A. Swami, Practical black-box attacks against machine learning, in *Proceedings of the 2017 ACM on Asia Conference on Computer and Communications Security*. ACM, pp. 506–519 (2017).

[11] S. Thys, W. Van Ranst and T. Goedemé, Fooling automated surveillance cameras: Adversarial patches to attack person detection (2019), *arXiv preprint* arXiv:1904.08653.

[12] S.-T. Chen, C. Cornelius, J. Martin and D. H. P. Chau, Shapeshifter: Robust physical adversarial attack on faster r-cnn object detector, in *Joint European Conference on Machine Learning and Knowledge Discovery in Databases*. Springer, pp. 52–68 (2018).

[13] J. Lu, H. Sibai, E. Fabry and D. Forsyth, No need to worry about adversarial examples in object detection in autonomous vehicles (2017), *arXiv preprint* arXiv:1707.03501.

[14] C. Xie, J. Wang, Z. Zhang, Y. Zhou, L. Xie and A. Yuille, Adversarial examples for semantic segmentation and object detection, in *Proceedings of the IEEE International Conference on Computer Vision*, pp. 1369–1378 (2017).

[15] J. H. Metzen, M. C. Kumar, T. Brox and V. Fischer, Universal adversarial perturbations against semantic image segmentation, in *2017 IEEE International Conference on Computer Vision (ICCV)*. IEEE, pp. 2774–2783 (2017).

[16] K. K. Nakka and M. Salzmann, Indirect local attacks for context-aware semantic segmentation networks, in *European Conference on Computer Vision*. Springer, pp. 611–628 (2020).

[17] N. Akhtar and A. Mian, Threat of adversarial attacks on deep learning in computer vision: A survey, *IEEE Access* **6**, pp. 14410–14430 (2018).

[18] A. Arnab, O. Miksik and P. H. Torr, On the robustness of semantic segmentation models to adversarial attacks, in *Proceedings of the IEEE Conference on Computer Vision and Pattern Recognition*, pp. 888–897 (2018).

[19] J. Al-afandi and H. András, Class retrieval of detected adversarial attacks, *Applied Sciences* **11**(14), p. 6438 (2021).

[20] Y. LeCun, The mnist database of handwritten digits (1998), http://yann.lecun.com/exdb/mnist/.

[21] J. Johnson, B. Hariharan, L. van der Maaten, L. Fei-Fei, C. L. Zitnick and R. Girshick, Clevr: A diagnostic dataset for compositional language and elementary visual reasoning, in *2017 IEEE Conference on Computer Vision and Pattern Recognition (CVPR)*. IEEE, pp. 1988–1997 (2017).

[22] O. Ronneberger, P. Fischer and T. Brox, U-net: Convolutional networks for biomedical image segmentation, in *International Conference on Medical Image Computing and Computer-Assisted Intervention*. Springer, pp. 234–241 (2015).

[23] K. He, G. Gkioxari, P. Dollár and R. Girshick, Mask r-cnn, in *Proceedings of the IEEE International Conference on Computer Vision*, pp. 2961–2969 (2017).

[24] Y. Ganin, E. Ustinova, H. Ajakan, P. Germain, H. Larochelle, F. Laviolette, M. Marchand and V. Lempitsky, Domain-adversarial training of neural networks, *The Journal of Machine Learning Research* **17**(1), pp. 2096–2030 (2016).

[25] E. Tzeng, K. Burns, K. Saenko and T. Darrell, Splat: Semantic pixel-level adaptation transforms for detection (2018), *arXiv preprint* arXiv:1812.00929.

[26] M. Cordts, M. Omran, S. Ramos, T. Rehfeld, M. Enzweiler, R. Benenson, U. Franke, S. Roth and B. Schiele, The cityscapes dataset for semantic urban scene understanding, in *Proceedings of the IEEE Conference on Computer Vision and Pattern Recognition (CVPR)* (2016).

[27] L.-C. Chen, G. Papandreou, F. Schroff and H. Adam, Rethinking atrous convolution for semantic image segmentation (2017), *arXiv preprint* arXiv:1706.05587.

[28] J. Hu, L. Shen and G. Sun, Squeeze-and-excitation networks, in *Proceedings of the IEEE Conference on Computer Vision and Pattern Recognition*, pp. 7132–7141 (2018).

[29] G. Montúfar, R. Pascanu, K. Cho and Y. Bengio, On the number of linear regions of deep neural networks (2014), *arXiv preprint* arXiv:1402.1869.

[30] H. Xiong, L. Huang, M. Yu, L. Liu, F. Zhu and L. Shao, On the number of linear regions of convolutional neural networks, in *International Conference on Machine Learning*. PMLR, pp. 10514–10523 (2020).

[31] J. Long, E. Shelhamer and T. Darrell, Fully convolutional networks for semantic segmentation, in *Proceedings of the IEEE Conference on Computer Vision and Pattern Recognition*, pp. 3431–3440 (2015).
[32] T. Serra, C. Tjandraatmadja and S. Ramalingam, Bounding and counting linear regions of deep neural networks, in *International Conference on Machine Learning*. PMLR, pp. 4558–4566 (2018).
[33] M. Trimmel, H. Petzka and C. Sminchisescu, Tropex: An algorithm for extracting linear terms in deep neural networks, in *International Conference on Learning Representations* (2021), https://openreview.net/forum?id=IqtonxWI0V3.

© 2025 World Scientific Publishing Company
https://doi.org/10.1142/9789811289125_0010

Chapter 10

Sentiment and Word Cloud Analysis of Tweets Related to COVID-19 Vaccines before, during, and after the Second Wave in India

Anmol Bansal, Arjun Choudhry, Anubhav Sharma, and Seba Susan*

*Department of Information Technology,
Delhi Technological University, Delhi, India*
seba_406@yahoo.in

Abstract

The COVID-19 pandemic caused by the severe acute respiratory syndrome coronavirus 2 (SARS-CoV-2) started in December 2019 in the city of Wuhan in China and rapidly spread worldwide in a matter of couple of months to cause a global health emergency. In the months following the initial spread of COVID-19, various countries and pharmaceutical companies ramped up their research to find a possible cure for the virus. In India, the first set of vaccines that were rolled out were Covaxin, developed by Bharat BioTech, and Covishield, jointly developed by Oxford and AstraZeneca. In this research work, we focus primarily on these two vaccines and evaluate their impact on social media, before, during, and after the second COVID-19 wave (caused due to the Delta variant of the Coronavirus) that affected India in mid-2021. Specifically, we gathered tweets in the duration of January 2021 to November 2021. We performed sentiment analysis using a pre-trained transformer model, and natural language processing using Word Clouds, to understand the dynamically changing public opinion on the vaccines as the deadly COVID-19 second wave unfolded, peaked and abated in the country, while the vaccination program was underway in the country.

10.1 Introduction

COVID-19 has spread worldwide for more than three years now and has claimed several million lives. To counter the coronavirus and develop herd immunity, vaccination drives were adopted in all countries. Since their introduction, vaccines have received mixed reviews, which vary over time as more awareness is spread among the masses [1]. Twitter (now X) was (is) a social media platform used by 368 million people worldwide, and it has played a significant role in building public opinion about the COVID-19 vaccines [2]. The COVID-19 vaccination drive in India commenced on 16th January 2021, and since then, more than 2.2 billion vaccine doses (including first, second, and booster doses) have been administered. Indians have primarily been inoculated with the Covaxin and Covishield vaccines, both of which were lauded as well as critically scrutinized on social networking platforms like Twitter or X [3]. Identifying the general sentiment of the public toward these vaccines helps us visualize their popularity, people's opinion of their efficacy, and whether there is an overwhelming inclination toward a particular vaccine [4–7]. Initially, only Covaxin and Covishield were administered in India, with paid options for Pfizer and Sputnik alternatives available only in recent months in select places. Hence, we restrict our purview only to Covaxin and Covishield which are the two popularly administered vaccines in India.

In this chapter, we present the sentiment analysis of Covaxin and Covishield-related posts before, during, and after the second wave of COVID-19 infections in India that was experienced between March 2021 and September 2021 and was attributed to the Delta mutant of the coronavirus. We used unlabeled COVID-19 vaccine-related posts downloaded from a large-scale dataset between January 2021 and November 2021 and employed transfer learning for classifying the unlabeled tweets. We used the pre-trained XLNet language model [8] for sentiment analysis that was fine-tuned on the sentiment-annotated *US Airline* tweets dataset [9]. This fine-tuned transformer model is applied to the unlabeled X vaccine dataset [10] to predict the respective sentiment for each post related to the vaccine Covaxin or Covishield. We further intend to track the perception of the masses since the arrival of the vaccines, through social media posts, and reflect on the reasons behind the dynamically evolving ideas of people. The construction of Word Clouds [11, 12] using Natural Language Processing (NLP) enables us to further evaluate the opinions expressed by people for each vaccine.

10.1.1 *Motivation*

The COVID-19 vaccination drive is continuing in full swing all across the globe, with countries ramping up orders and production of booster vaccines. In countries like India, where more than one type of COVID-19 vaccine is available to the public, people's opinion of each vaccine brand is affected by social media as well as external means. To determine the level of acceptance for these vaccines in the eyes of the general public, it is important to detect the sentiments portrayed by people discussing them on social media platforms and thereafter evaluate whether each vaccine is accepted positively or negatively by the people and how these views vary over a period of time.

Researchers working in the field of social media analysis typically use posts from social media platforms as a source of data. Twitter, one of the most popular social networking platforms with millions of users all over the world, was one of the largest sources of textual data used for NLP tasks. To depict the sentiments of posts posted by people on social media, posts extracted from Twitter using the Tweepy Python library were analyzed to determine the opinion of people on each vaccine before, during, and after the second wave of COVID-19 infections in India, ranging over a span of 11 months from January 2021 to November 2021. We further implement Word Clouds using NLP to qualitatively judge the text in the positive, neutral, and negative sentiment posts during the same period.

In this chapter, we analyze how public opinion varies over time with respect to each vaccine as well as the vaccination drive in general. To compare the accuracy of the predicted sentiments of posts, we use a variety of supervised and non-supervised models; we first evaluate the model performance on an annotated social media dataset for sentiment analysis, namely the sentiment-annotated *US Airline* tweets dataset. Specifically, we evaluate the performance of supervised models like Bidirectional LSTM (Bi-LSTM), BERT, and XLNet, fine-tuned on the sentiment-annotated *US Airline* tweets dataset, and analyze their performance against unsupervised models like VADER and TextBlob for sentiment analysis.

10.1.2 *Problem statement*

In this chapter, we analyze how people's opinion of each vaccine is affected over the span of 11 months starting from January 2021 to November 2021. This chapter aims to analyze the general sentiments of people toward Covaxin and Covishield before, during, and after the second wave of COVID-19 infections in India and demonstrates how social media has captured people's inclination toward a particular vaccine. We analyze and evaluate the patterns observed in the variation of public opinion on various COVID-19 vaccines, and the vaccination drive in general, using transfer learning-based supervised sentiment analysis models, unsupervised sentiment analysis models, and Word Clouds.

Some of the specific objectives covered in this chapter are as follows:

1. analyzing the performance of different supervised and unsupervised learning approaches for sentiment analysis in order to identify the best approach for sentiment annotation of the unlabeled COVID-19 vaccine-related posts,
2. evaluating the performance of a fine-tuned XLNet transformer model for sentiment analysis of COVID-19 vaccine-related posts over other transformer-based models, such as BERT, for a more detailed insight into the use of transfer learning for sentiment analysis,
3. presenting an overview on the reception of the two vaccines administered in India, Covaxin and Covishield, based on the posts posted by people on Twitter, how people's opinion varies over time, and possible factors for variations in sentiments portrayed by the posts,

4. implementing topic analysis for sentiment annotated posts by constructing and analyzing Word Clouds for qualitatively evaluating the sentiments observed between January 2021 and November 2021, in an attempt to understand the reason behind the sentiment fluctuations.

10.1.3 *Related works*

Sentiment analysis, which has been well researched in the field of natural language processing, has seen widespread use in a variety of domains and tasks. It is widely used in social media analysis for tasks like stock movement prediction [13, 14], fake news detection [15, 16], and study of mental health [17] in order to find the emotional tone of the text in the social media posts.

There have been several initiatives by researchers to assess the global impact of COVID-19 using social media analytics, specifically by analyzing social media posts using supervised and unsupervised approaches as well as natural language processing techniques. The authors in their previous research explored transformer models for COVID-19 vaccine sentiment analysis [18] and proved the futility of text oversampling techniques [19] for countering the data inadequacy problem. Supervised learning requires adequate training data which is lacking for vaccine-related information on social media. Dynamic topic modeling was proved useful in [20] for qualitatively understanding the topmost topics under discussion in certain time frames. Word Clouds were employed by [21, 22] to understand the topmost topics under discussion during the course of the pandemic.

Several supervised and unsupervised models have been proposed in the literature for sentiment analysis of COVID-19 posts on social media. We discuss a few of these next. Prabucki (2021) evaluated three deep learning models, namely LSTM, Bi-LSTM, and 1D-CNN, trained on the Twitter Sentiment Extraction dataset for evaluating the sentiments associated with various COVID-19 vaccines [23]. The Bi-LSTM model outperformed the other two models on the Twitter Sentiment Extraction dataset and was further used for classifying the COVID-19 Vaccination Tweets dataset. Na *et al.* used unsupervised sentiment analysis for evaluating the sentiments associated with various COVID-19 vaccines [24]; specifically, they adopted the

VADER framework [25] for sentiment analysis. They further evaluated the impact of the influence of celebrities and showed that posts by celebrities for or against COVID-19 vaccines led to significant shifts in the opinion of the people toward the vaccines. Marcec and Likic [7] performed sentiment analysis on a custom-made dataset consisting of posts mentioning Oxford/AstraZeneca, Moderna, and Pfizer vaccines. They used the AFINN lexicon [26] for sentiment analysis and found that the sentiments related to posts on Moderna and Pfizer vaccines were mostly stable and positive over a four-month span. The sentiments of posts on Covishield/AstraZeneca, on the other hand, kept dropping to more negative values.

10.2 Methodology

This section starts with the basic concepts used in our approach including a review of the sentiment analysis task and the XLNet transformer. This is followed by a detailed description of the proposed method that involves transfer learning for sentiment analysis of unlabeled COVID-19 vaccine-related posts. The basic process flow of our experiments is described in detail.

10.2.1 *Sentiment analysis of social media posts*

Sentiment analysis, a subset of opinion mining, is a natural language processing approach for the contextual detection of the emotional tone behind a piece of text and is useful for identifying, extracting, and understanding the emotions and opinions expressed by the writer. Generally, a piece of text is said to have one of the three base sentiments: positive, neutral, or negative. Sentiment analysis has been popularly applied for the evaluation of social media posts, especially on Twitter due to the massive flow of information on the platform [27–31].

10.2.2 *XLNet*

XLNet [8] was proposed by Yang *et al.* It is a generalized auto-regressive pre-trained model, built based on the Transformer-XL model. Unlike BERT, which is an auto-encoder model, XLNet, being an auto-regressive model, completely removes the need to mask the data and denoise during fine-tuning on a downstream task. An auto-regressive model is inherently flawed due to its unidirectional nature. To counter this issue, Yang *et al.* proposed the Permutation Language Modeling task, in which tokens are predicted using the preceding tokens, however, each token, instead of being taken as placed in the sequence, is randomly shuffled. Thereby, multiple orders of the same sequence are used to train the XLNet model, and each sequence occurs in all possible permutations of its tokens. The XLNet model further inculcates the two key approaches from Transformer-XL, namely Relative Positional Embeddings and Recurrence Mechanism. Further, the authors used two-stream self-attention to keep the knowledge of the position of the token it is predicting.

10.2.3 COVID-19 vaccine sentiment analysis before, during, and after second wave in India by transfer learning using XLNet

Here, we briefly describe our recent work [3] on sentiment analysis of vaccine-related posts during the second wave in India and also discuss the extensions to our project to create an enhanced view of the effect the pandemic had on public opinion about vaccines. This chapter presents a more detailed examination of an extended time framework, January 2021 to November 2021, as compared to [3] which was constrained to March–September 2021 time frame when the second wave was at its peak. We top up our current experiments with a topic analysis segment, in which we implement Word Clouds for sentiment-annotated posts using NLP techniques. The current work is motivated by the need for testing the public sentiments and understanding the topics that are associated with different sentiments related to the two vaccines popularly administered in India: Covishield and Covaxin. The majority of the Indian population has taken both doses of either of the two vaccines to date. Covishield is the Oxford AstraZeneca vaccine, while Covaxin is an indigenous Indian vaccine manufactured by Bharat BioTech and approved for use in India.

10.2.3.1 Dataset preparation

The COVID-19 All Vaccines Tweets dataset [10] contains posts collected using the Tweepy Python library. This dataset is provided on

Kaggle and is one of the most popular datasets for sentiment analysis of COVID-19 vaccine-related posts. It contains posts related to seven different vaccines and has been updated till 30th November 2021 with a total of 223,288 posts. We separated out the posts pertaining to either Covishield (also referred to as AstraZeneca) or Covaxin in the duration of 1st January 2021 to 30th November 2021. We have not considered the posts having a mention of both the vaccines, for the sake of simplicity. The total number of Covaxin posts segregated by this procedure was 65,164 while that of Covishield posts was 11,190. The posts were cleaned of URLs and links to websites, and all sentences were converted to lowercase. Due to the absence of labels in the COVID-19 All Vaccines Tweets dataset, a separate dataset labeled with sentiment annotations — the *US Airline* tweets dataset [9] — was used to fine-tune a transformer model loaded with XLNet pre-trained weights. The fine-tuned model was then used to predict the sentiments for the unlabeled vaccine-related posts.

10.2.3.2 *Experiments*

We obtained posts before, during, and after the second COVID-19 wave in India in the duration of 1st January 2021 to 30th November 2021 (spanning 11 months) from the large-scale COVID-19 All Vaccines Tweets dataset. These posts are devoid of sentiment annotations, which is an obstacle to supervised learning. In order to classify the unlabeled posts into positive, negative, and neutral sentiments, we perform transfer learning by fine-tuning the pre-trained XL-Net transformer model on the *US Airline* tweets dataset [9] which is one of the benchmark datasets for post-based sentiment analysis. XLNet is an auto-regressive pre-trained transformer model with multi-head self-attention that has achieved high accuracies for text classification tasks [32,33] as compared to the LSTM-based encoder–decoder models [34]. XLNet is an extension of the transformer-XL model [35,36] that has outperformed the simple transformer for different text classification and generation tasks [37]. XLNet introduced a modified language model training objective which learns distributions for all permutations of sequence tokens.

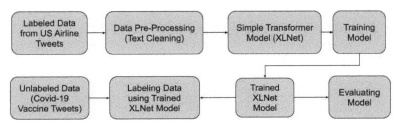

Fig. 10.1. Framework for sentiment analysis on *US Airline* tweets to facilitate sentiment annotation of the COVID-19 All Vaccine Tweets dataset using the trained XLNet model.

The training process for the transformer model is depicted in the process flow in Figure 10.1. The XLNet base model has 12 layers, 768 hidden states, 12 attention heads, and 110 million parameters. We fine-tune the pre-trained XLNet model on the sentiment-annotated *US Airline* tweets dataset as per the guidelines in our recent work [3]. We use the 'xlnet-base-cased' model with the AdamW optimizer, trained on 11,712 samples of the *US Airline* tweets dataset over 5 epochs with a batch size of 16.

The overall process is shown as two separate phases in Figures 10.2(a) and 10.2(b), respectively. In the first phase, the XLNet model is fine-tuned on the *US Airline* tweets that are sentiment annotated. A three-fold cross-validation is performed for training the XLNet model on the *US Airline* tweets. In the second phase of our experiments, shown in Figure 10.2(b), the fine-tuned XLNet model is used to classify the unlabeled COVID-19 vaccine-related posts pertaining to the two vaccines, Covishield and Covaxin, that were popularly administered in India.

The posts so labeled are now analyzed over time using natural language processing techniques to determine if the overall attitude toward vaccines has undergone a change and if there is a boost in the general positivity of social media users toward COVID-19 vaccines during the course of the pandemic. We present Word Clouds of high-frequency unigrams extracted from different time windows as evidence. This work is motivated by the need for dynamically monitoring the public sentiments related to the COVID-19 vaccines during the course of the vaccination program while an infection wave started, peaked, and abated in the country.

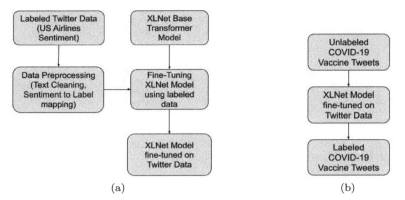

Fig. 10.2. Process flow: (a) fine-tuning the pre-trained XLNet model on the sentiment-annotated *US Airline* tweets; (b) classification of COVID-19 vaccine-related posts using the fine-tuned XLNet model.

10.3 Experimental Setup

This section contains a description of the two datasets used in our experiments followed by the data preprocessing steps.

10.3.1 *Datasets used*

We have used two datasets in this work:

1. *US Airline* tweets dataset [9],
2. COVID-19 All Vaccines Tweets dataset [10].

The sentiment-annotated *US Airline* tweets dataset [9] is contributed by CrowdFlower's Data for Everyone library and contains posts and their respective labels for sentiment analysis. We use this dataset to train our supervised learning models, which are then used to analyze the sentiments of the COVID-19 Vaccination Tweets Dataset.

The COVID-19 All Vaccines Tweets dataset available on Kaggle [10] contains posts pertaining to almost all vaccine brands, such as Moderna, Pfizer, Sinopharm, Sputnik-V, Covaxin, and AstraZeneca/ Covishield. It is contributed by Gabriel Preda on Kaggle and contains a total of 223,288 posts. The data were collected using the Tweepy

Table 10.1. COVID-19 All Vaccines Tweets dataset breakdown.

Column	Type	Unique Values
id	int64	223,288
user_name	string	84,274
user_location	string	25,123
user_description	string	82,704
user_created	string	85,467
user_followers	int64	22,114
user_friends	int64	8,124
user_favourites	int64	38,200
user_verified	boolean	2
date	string	217,222
text	string	221,464
hashtags	string	59,085
source	string	371
retweets	int64	447
favourites	int64	981
is_retweet	boolean	2

Python library. For each vaccine, relevant search terms were used (most frequently used in the social media platform to refer to the respective vaccine). The dataset has 16 columns (id, user_name, user_location, user_description, user_created, user_followers, user_friends, user_favourites, user_verified, date, text, hashtags, source, retweets, favorites, and is_retweet). The details of the dataset are shown in Table 10.1.

10.3.2 *Data preprocessing*

The social media platform data contain a large number of mentions and hashtags, which serve minimal purpose for our sentiment analysis task. Therefore, we preprocessed both datasets to get clean text data for sentiment analysis by following the following steps:

1. removing email addresses, mentions of others, and hashtags,
2. removing links to other webpages,
3. removing extra spaces in the text.

Table 10.2. COVID-19 Vaccination Tweets breakdown for Covaxin and Covishield.

Vaccine	Number of Posts
Covaxin	65,164
Covishield	11,190

We limited our evaluation to the Covishield and Covaxin vaccines and therefore extracted datapoints specifically for these vaccines. The distribution of the number of posts for the two vaccines is highlighted in Table 10.2.

After tokenization, all words are converted to lowercase and the stop words are removed from the posts. URLs and email ids in the posts are also removed. Then all the words are stemmed (Snowball Stemmer, in our case). The post texts belonging to a particular time window are converted into the Bag of Words (BoW) dictionary by extracting words (tokenizing) and counting the number of times they occur (term frequency). The BoW model has been used successfully for various text classification experiments [38, 39]. The BoW dictionary is then filtered by removing words that are less frequent (less than 15 in our case). The BoW features are used to construct the Word Clouds.

10.4 Results and Discussions

We evaluated the performance of five different sentiment analysis models trained on the sentiment-annotated *US Airline* tweets to depict the best fit for the social media platform data. The results for each model in terms of accuracy and F1 score are shown in Table 10.3.

The unsupervised learning models VADER and TextBlob are significantly outperformed by the supervised Bi-LSTM, BERT, and XLNet models as per the classification scores in Table 10.3. The notably superior performance of the supervised models can be attributed to the significant number of neutral sentiment posts in the *US Airline* dataset, which are majorly misinterpreted by VADER and TextBlob due to their requirement of a sentiment score of zero

Table 10.3. Performance comparison for sentiment analysis of the US Airline dataset.

Method	Accuracy (%)	F1-score
VADER	45.3	0.536
TextBlob	44.4	0.446
Bi-LSTM	81.1	0.801
BERT	84.2	0.843
XLNet	85.8	0.856

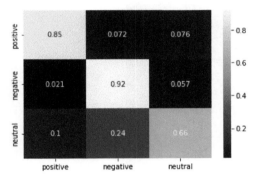

Fig. 10.3. Confusion matrix of the XLNet model trained on the US Airline tweets dataset.

for a post to be labeled with neutral sentiment. Among the supervised learning models, it is noted that the performance of Bi-LSTM falls short of the transformer-based language models. BERT noticeably outperforms the unsupervised learning models as well as the Bi-LSTM model. However, it is unable to match the performance of XLNet, primarily due to the fact that XLNet does not require denoising the data while fine-tuning downstream tasks.

It was noted that the removal of stopwords and punctuations lowered the accuracy of the XLNet model. The model had some difficulty in distinguishing between the negative and neutral sentiments, but the precision and recall for both positive and negative posts were high as observed from the confusion matrix in Figure 10.3. The XLNet model was able to correctly classify 85% of the positive posts and 92% of the negative posts.

The transfer-learned XLNet model is now used to classify COVID-19 vaccine-related posts into three classes: positive, negative, and neutral. After annotating the posts for Covaxin and Covishield with their sentiments using the fine-tuned XLNet model, we observed that a significant chunk of the posts was labeled as neutral, possibly due to their informational content, instead of portraying the writer's individual views on a vaccine. As observed from the bar charts in Figure 10.4, 12.25% of the total posts (=65,164) for Covaxin are negative and 7.3% are positive, while the remaining

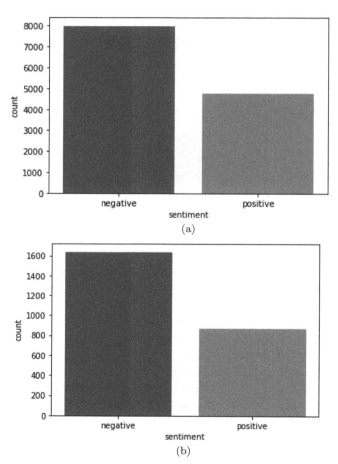

Fig. 10.4. Distribution of negative and positive posts for (a) Covaxin and (b) Covishield during the second wave in India.

are neutral. The Covishield posts (=11,190) also showed a similar trend with 14.6% negative posts and 7.7% positive posts, and the remaining posts being neutral. The neutral class is not shown in Figure 10.4 since neutral posts are mostly informational posts and have no direct bearing on the public sentiment. It should be noted that a larger training corpus would further help in increasing the performance of supervised learning models for social media analysis.

The negative sentiment exceeds the positive sentiment for both vaccines during the second wave from March 2021 to September 2021 due to apprehensions regarding the vaccination process during the course of the second wave of the pandemic in India.

People exhibit their sentiments about an entity by using adjectives to specify what they feel about the entity. By analyzing the Word Clouds (consisting mainly of adjectives), we can find the overall sentiments related to the vaccines. The Word Clouds are shown plotted in Figures 10.5, 10.6, and 10.7 for the positive, negative, and neutral posts pertaining to Covaxin, and in Figures 10.8, 10.9, and 10.10 for the positive, negative, and neutral posts pertaining to Covishield.

The Word Clouds for Covaxin positive sentiments in Figure 10.5 highlight words such as *good*, *best*, *proud*, and *Indian*, which shows that people are proud that Covaxin is an Indian vaccine. *First* is more prevalent from January 2021 to May 2021, possibly due to demand for the first dose of vaccines. From June 2021 to November 2021, demand for the second dose of vaccines was increasing and hence the mention of the word *second*.

The Word Clouds for Covaxin negative sentiments in Figure 10.6 depict words such as *available*, *first*, *traditional*, and *due*, which shows that people are flustered about the lack of availability of the vaccine. The use of the words *inactive* and *traditional* further points out people's apprehension about the fact that Covaxin is made from the inactive Coronavirus and is more akin to traditional vaccines. The frequent use of the word *international* in the graph from June 2021 to November 2021 further calls into account people's frustration regarding the lack of international recognition given to Covaxin.

The Word Clouds for Covaxin neutral sentiments in Figure 10.7 depict words such as *inactivated*, *nationwide*, *clinical*, *diagnostic*, and *biological*, which shows that these posts contain significantly more

Fig. 10.5. Word Clouds for Covaxin-positive posts from (a) January 2021 to May 2021; (b) June 2021 to November 2021.

Sentiment and Word Cloud Analysis 245

(a)

(b)

Fig. 10.6. Word Clouds for Covaxin negative posts from (a) January 2021 to May 2021; (b) June 2021 to November 2021.

(a)

(b)

Fig. 10.7. Word Clouds for Covaxin neutral posts from (a) January 2021 to May 2021; (b) June 2021 to November 2021.

nouns, indicating more factual information as compared to positive and negative posts. The posts are mostly informative rather than used for conveying one's opinion on the social media platform.

The Word Clouds for Covishield in Figures 10.8, 10.9, and 10.10 follow trends similar to Covaxin, with *first* being present in all the graphs, while *second* is present in graphs from June 2021 to November 2021. The negative Word Clouds from June 2021 to November 2021 further show widespread use of the words *fatal, adverse, dangerous,* and *cardiac* with people calling into question the side effects of Covishield leading to cardiac arrests in people.

The sentiment line graphs for Covaxin are shown plotted in Figures 10.11 and 10.12. The posts for Covaxin from January 2021 to May 2021 in Figure 10.11 show that negative posts are noticeably higher in number than positive posts. This is likely due to people's hesitance in taking the vaccine, lack of international recognition for Covaxin, and low availability of doses, as noticed from the patterns in the Word Clouds in Figures 10.5, 10.6, and 10.7.

The posts from June 2021 to November 2021 in Figure 10.12 continue this trend with more negative than positive posts, but positive posts are very close behind their counterpart. This is the result of low availability of vaccines during the second wave. The massive spike in the positive posts in October 2021 and November 2021 can be attributed to international acceptance of Covaxin, WHO approval of Covaxin as a vaccine for emergency use, and reports stating the greater efficacy of Covaxin against the Delta variant, as compared to other vaccines.

The posts for Covishield from January 2021 to May 2021 in Figure 10.13 show that negative posts are noticeably higher in number than positive posts. This is likely due to people's hesitance in taking the vaccine and low availability of doses, similar to what was observed for Covaxin.

The posts from June 2021 to November 2021 in Figure 10.14 continue this trend with more negative than positive posts. This is also the result of low availability of vaccines during the second wave. Since the number of posts for Covishield is low, the resultant graph is more likely to be skewed and less conclusive than the one for Covaxin.

(a)

(b)

Fig. 10.8. Word Clouds for Covishield positive posts from (a) January 2021 to May 2021; (b) June 2021 to November 2021.

Sentiment and Word Cloud Analysis 249

(a)

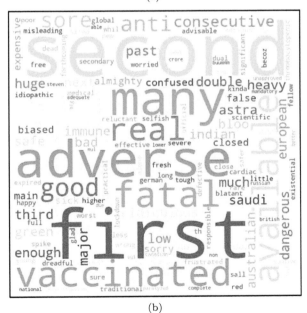

(b)

Fig. 10.9. Word Clouds for Covishield negative posts from (a) January 2021 to May 2021; (b) June 2021 to November 2021.

Fig. 10.10. Word Clouds for Covishield neutral posts from (a) January 2021 to May 2021; (b) June 2021 to November 2021.

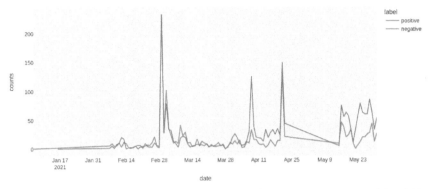

Fig. 10.11. Line graph for Covaxin sentiments from January 2021 to May 2021.

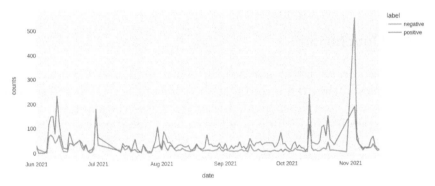

Fig. 10.12. Line graph for Covaxin sentiments from June 2021 to November 2021.

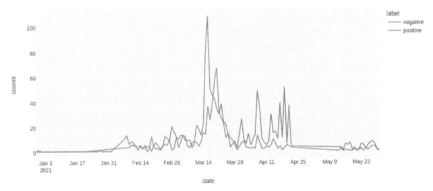

Fig. 10.13. Line graph for Covishield sentiments from January 2021 to May 2021.

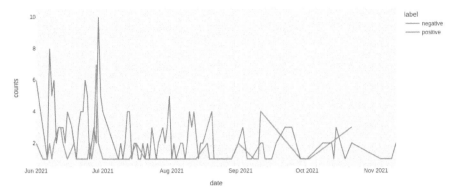

Fig. 10.14. Line graph for Covishield sentiments from June 2021 to November 2021.

10.5 Conclusion

In this chapter, we have analyzed the sentiments associated with the two vaccines authorized in India at the time of the COVID-19 second wave: Covaxin and Covishield. In future work, the analysis can be further expanded to include other vaccines, including the following:

- Pfizer,
- Sinopharm,
- Sinovac,
- Moderna,
- Sputnik-V.

Analyzing other vaccines can give us an insight into their acceptance levels across the globe. More in-depth comparative analysis can be done on vaccines based on their success levels in different countries, as well as how they are perceived and accepted in different parts of the world. Analyzing the reactions of people in different countries can help us understand the trend followed by people with respect to vaccines' development.

In this chapter, only those posts which mention only one of the two vaccines are considered. However, there are many posts which contain mentions of multiple vaccines that were not considered due to their ambiguous nature. A Named Entity Recognition-based sentiment analysis model [40] can be developed to address this issue in order to determine the key entities in the text prior to sentiment analysis. This will help not only to generate a larger dataset but also in accurate prediction of the sentiments corresponding to different vaccines.

References

[1] N. Puri, E. A. Coomes, H. Haghbayan and K. Gunaratne, Social media and vaccine hesitancy: New updates for the era of COVID-19 and globalized infectious diseases, *Human Vaccines & Immunotherapeutics* **16**(11), pp. 2586–2593 (2020).

[2] S. Yousefinaghani, R. Dara, S. Mubareka, A. Papadopoulos and S. Sharif, An analysis of Covid-19 vaccine sentiments and opinions on X, *International Journal of Infectious Diseases* **108**, pp. 256–262 (2021). https://doi.org/10.1016/j.ijid.2021.05.059.

[3] A. Bansal, S. Susan, A. Choudhry and A. Sharma, Covid-19 vaccine sentiment analysis during second wave in india by transfer learning using XLNet, in *Pattern Recognition and Artificial Intelligence: 3rd International Conference, ICPRAI 2022*, Paris, France, June 1–3, 2022, Proceedings, Part II. Springer International Publishing, Cham, pp. 443–454 (2022).

[4] A. Hussain, A. Tahir, Z. Hussain, Z. Sheikh, M. Gogate, K. Dashtipour, A. Ali and A. Sheikh, Artificial intelligence–enabled analysis of public attitudes on facebook and twitter toward covid-19 vaccines in the United Kingdom and the United States: Observational study, *Journal of Medical Internet Research* **23**(4), p. e26627 (2021).

[5] C. A. Melton, O. A. Olusanya, N. Ammar and A. Shaban-Nejad, Public sentiment analysis and topic modeling regarding COVID-19 vaccines on the Reddit social media platform: A call to action for strengthening vaccine confidence, *Journal of Infection and Public Health* **14**(10), pp. 1505–1512 (2021).

[6] F. M. Shamrat, S. Chakraborty, M. M. Imran, J. N. Muna, M. Billah, P. Das and M. Rahman, Sentiment analysis on twitter tweets about COVID-19 vaccines using NLP and supervised KNN classification algorithm, *Indonesian Journal of Electrical Engineering and Computer Science* **23**, pp. 463–470 (2021). doi: 10.11591/ijeecs.v23.i1.pp463-470.

[7] R. Marcec and R. Likic, Using X for sentiment analysis towards AstraZeneca/Oxford, Pfizer/BioNTech and Moderna COVID-19 vaccines, *Postgraduate Medical Journal* (2021), https://doi.org/10.1136/postgradmedj-2021-140685.

[8] Z. Yang, Z. Dai, Y. Yang, J. Carbonell, R. R. Salakhutdinov and Q. V. Le, XLNet: Generalized autoregressive pretraining for language understanding, in *Advances in Neural Information Processing Systems*, Vol. 32 (2019), https://proceedings.neurips.cc/paper/2019/file/dc6a7e655d7e5840e66733e9ee67cc69-Paper.pdf.

[9] F. Eight, X US Airline Sentiment (4), Kaggle (2016), https://www.kaggle.com/crowdflower/twitter-airline-sentiment.

[10] G. Preda, COVID-19 All Vaccine Tweets (110), Kaggle (2021), https://www.kaggle.com/gpreda/all-covid19-vaccines-tweets.

[11] A. I. Kabir, R. Karim, S. Newaz and M. I. Hossain, The power of social media analytics: Text analytics based on sentiment analysis and word clouds on R, *Informatica Economica* **22**(1), pp. 25–38 (2018).

[12] S. Giannoulakis and N. Tsapatsoulis, Topic identification via human interpretation of word clouds: The case of instagram hashtags, in *Artificial Intelligence Applications and Innovations: 17th IFIP WG 12.5 International Conference, AIAI 2021*, Hersonissos, Crete, Greece, June 25–27, 2021, Proceedings 17. Springer International Publishing, pp. 283–294 (2021).

[13] T. H. Nguyen, K. Shirai and J. Velcin, Sentiment analysis on social media for stock movement prediction, *Expert Systems with Applications* **42**(24), pp. 9603–9611 (2015), https://doi.org/10.1016/j.eswa.2015.07.052.

[14] A. Derakhshan and H. Beigy, Sentiment analysis on stock social media for stock price movement prediction, *Engineering Applications of Artificial Intelligence* **85**, pp. 569–578 (2019), https://doi.org/10.1016/j.engappai.2019.07.002.

[15] S. Kula, M. Choraś, R. Kozik, P. Ksieniewicz and M. Woźniak, Sentiment analysis for fake news detection by means of neural networks, in V. V. Krzhizhanovskaya, G. Závodszky, M. H. Lees, J. J. Dongarra, P. M. A. Sloot, S. Brissos and J. Teixeira (eds.), *Computational Science — ICCS 2020*. Springer International Publishing, pp. 653–666 (2020).

[16] M. A. Alonso, D. Vilares, C. Gómez-Rodríguez and J. Vilares, Sentiment analysis for fake news detection, *Electronics* **10**(11) (2021), https://doi.org/10.3390/electronics10111348.

[17] Z. Chen and M. Sokolova, Sentiment analysis of the COVID-related r/depression posts (2021), arXiv: abs/2108.06215.

[18] A. Bansal, A. Choudhry, A. Sharma and S. Susan, Adaptation of domain-specific transformer models with text oversampling for sentiment analysis of social media posts on COVID-19 vaccine, *Computer Science* **24**(2) (2023).

[19] A. Choudhry, S. Susan, A. Bansal and A. Sharma, TLMOTE: A topic-based language modelling approach for text oversampling, in *The International FLAIRS Conference Proceedings*, Vol. 35 (2022).

[20] A. Sharma, S. Susan, A. Bansal and A. Choudhry, Dynamic topic modeling of Covid-19 vaccine-related tweets, in *2022 the 5th International Conference on Data Storage and Data Engineering*, pp. 79–84 (2022).

[21] M. O. Lwin, J. Lu, A. Sheldenkar, P. J. Schulz, W. Shin, R. Gupta and Y. Yang, Global sentiments surrounding the COVID-19 pandemic on X: Analysis of X trends, *JMIR Public Health and Surveillance* **6**(2), p. e19447 (2020).

[22] D. Gandasari and D. Dwidienawati, Content analysis of social and economic issues in Indonesia during the COVID-19 pandemic, *Heliyon* **6**(11), p. e05599 (2020).

[23] T. Prabucki, Sentiment analysis of SARS-CoV-2 vaccination tweets using deep neural networks. figshare (2021), https://doi.org/10.6084/m9.figshare.14365292.v1.

[24] T. Na, W. Cheng, D. Li, W. Lu and H. Li, Insight from NLP analysis: COVID-19 vaccines sentiments on social media (2021), arXiv: abs/2106.04081.

[25] C. J. Hutto and E. Gilbert, VADER: A parsimonious rule-based model for sentiment analysis of social media Text, in *Proceedings of the 8th International Conference on Weblogs and Social Media, ICWSM 2014* (2015).

[26] F. Koto and M. Adriani, A comparative study on twitter sentiment analysis: Which features are good? in *Natural Language Processing and Information Systems: 20th International Conference on Applications of Natural Language to Information Systems, NLDB 2015, Passau, Germany, June 17–19, 2015, Proceedings 20*. Springer International Publishing, pp. 453–457 (2015).

[27] L. I. Tan, W. S. Phang, K. O. Chin and A. Patricia, Rule-based sentiment analysis for financial news, in *2015 IEEE International Conference on Systems, Man, and Cybernetics* pp. 1601–1606 (2015), https://doi.org/10.1109/SMC.2015.283.

[28] J. Singh, G. Singh and R. Singh, Optimization of sentiment analysis using machine learning classifiers, *Human-centric Computing and Information Sciences* **7**(1), p. 32 (2017). doi:10.1186/s13673-017-0116-3.

[29] M. Ghosh, K. Gupta and S. Susan, Aspect-based unsupervised negative sentiment analysis, in *Intelligent Data Communication Technologies and Internet of Things*. Springer, Singapore, pp. 335–344 (2021).

[30] F. Å. Nielsen, A new ANEW: Evaluation of a word list for sentiment analysis in microblogs (2011), arXiv: abs/1103.2903.

[31] S. Vashishtha and Susan, S. Neuro-fuzzy network incorporating multiple lexicons for social sentiment analysis, *Soft Computing*, **26**(9), pp. 4487–4507 (2022).

[32] X. He and V. O. K. Li., Show me how to revise: Improving lexically constrained sentence generation with XLNet, in *Proceedings of the AAAI Conference on Artificial Intelligence*, Vol. 35, no. 14, pp. 12989–12997 (2021).

[33] A. Sweidan, N. El-Bendary and H. Al-Feel, Sentence-level aspect-based sentiment analysis for classifying adverse drug reactions (ADRs) using hybrid ontology-XLNet transfer learning, *IEEE Access* **1** (2021), https://doi.org/10.1109/ACCESS.2021.3091394.

[34] R. Goel, S. Vashisht, A. Dhanda and S. Susan, An empathetic conversational agent with attentional mechanism, in *2021 International Conference on Computer Communication and Informatics (ICCCI)*. IEEE, pp. 1–4 (2021).

[35] Z. Dai, Z. Yang, Y. Yang, J. G. Carbonell, Q. Le and R. Salakhutdinov, Transformer-XL: Attentive language models beyond a fixed-length context, in *Proceedings of the 57th Annual Meeting of the Association for Computational Linguistics*, pp. 2978–2988 (2019).

[36] R. Goel, S. Susan, S. Vashisht and A. Dhanda, Emotion-aware transformer encoder for empathetic dialogue generation, in *2021 9th International Conference on Affective Computing and Intelligent Interaction Workshops and Demos (ACIIW)*, September. IEEE, pp. 1–6 (2021).

[37] W. Rahman, M. K. Hasan, S. Lee, A. Zadeh, C. Mao, L. P. Morency and E. Hoque, Integrating multimodal information in large pretrained transformers, in *Proceedings of the Conference. Association for Computational Linguistics. Meeting*, July, Vol. 2020. NIH Public Access, p. 2359 (2020).

[38] M. Sharma, G. Choudhary and S. Susan, Resume classification using elite bag-of-words approach, in *2023 5th International Conference on Smart Systems and Inventive Technology (ICSSIT)*. IEEE, pp. 1409–1413 (2023).
[39] S. Susan and J. Keshari, Finding significant keywords for document databases by two-phase maximum entropy partitioning, *Pattern Recognition Letters* **125**, pp. 195–205 (2019).
[40] L. Nemes and A. Kiss, Information extraction and named entity recognition supported social media sentiment analysis during the COVID-19 pandemic, *Applied Sciences* **11**(22), p. 11017 (2021).

Chapter 11

Reinforcement Learning and Sequential QAP-Based Graph Matching for Semantic Segmentation of Images

Jérémy Chopin[*,¶], Jean-Baptiste Fasquel[*,‖],
Harold Mouchère[†,**], Rozenn Dahyot[‡,††] and
Isabelle Bloch[§,‡‡]

[*]*LARIS, Université d'Angers, Angers, France*
[†]*LS2N, Universite de Nantes, CNRS UMR, Nantes 6004, France*
[‡]*Department of Computer Science, Maynooth University, Maynooth, Ireland*
[§]*Sorbonne Université, CNRS, LIP6, Paris, France*
[¶]*jeremy.chopin@univ-angers.fr*
[‖]*jean-baptiste.fasquel@univ-angers.fr*
[**]*harold.mouchere@univ-nantes.fr*
[††]*Rozenn.Dahyot@mu.ie*
[‡‡]*isabelle.bloch@sorbonne-universite.fr*

Abstract

This chapter addresses the fundamental task of semantic image analysis by exploiting structural information (spatial relationships between image regions). We propose to combine a deep neural network (DNN) with graph matching (formulated as a quadratic assignment problem (QAP)) where graphs encode efficiently structural information related to regions segmented by the DNN. Our novel approach solves the QAP sequentially for matching graphs, in the context of image semantic segmentation, where the optimal sequence for graph matching is conveniently defined using reinforcement learning (RL) based on the region membership probabilities produced by the DNN and their structural relationships. Our RL-based strategy for solving QAP sequentially allows us to significantly reduce the combinatorial complexity for graph matching. Two experiments are performed on two public datasets dedicated respectively to the semantic segmentation of face images and sub-cortical region of the brain. Results show that the proposed RL-based ordering performs better than using a random ordering, especially when using DNNs that have been trained on a limited number of samples. The open-source code and data are shared with the community.[a]

11.1 Introduction

Semantic segmentation is a fundamental but challenging task in computer vision, often managed using deep neural networks, such as U-Net [1]. Structural information [2,3], such as spatial relationships, is not explicitly used in such networks, although some recent works aim at exploiting it, e.g., CRF-based approaches [4] and CNN-based semantic segmentation followed by inexact graph matching [5].

In this chapter, we focus likewise on graph-based approaches exploiting relationships observed at high semantic levels in annotated training images or provided by qualitative descriptions of the scene content [3]. In this context, graph vertices and edges encode regions and spatial relationships produced by a segmentation network and observed in annotated training images, leading to an inexact graph-matching problem, expressed classically as a quadratic assignment problem (QAP) [6]. Note that some recent approaches solve graph matching with machine learning (e.g., graph neural networks [7]). Although promising for many application domains [8], large and representative training datasets of annotated graphs are required. Another difficulty is the definition of the appropriate architecture, and the management of both vertex and edge information, while edge features (related to relationships between regions) are often ignored [8].

One of the main drawbacks of QAP-based graph matching lies in its highly combinatorial nature [6]. In this context, our proposal is

---

[a]https://github.com/Jeremy-Chopin/APACoSI/.

to solve it in a sequential manner, where vertices are progressively matched in order to reduce the complexity. This means that the semantic image analysis is done progressively: First, identified regions are used to discover next ones [9, 10] (this is close to the notion of seeded graph matching [11]). The difficulty is to learn the optimal segmentation/graph-matching order, to ensure that all regions are finally recovered. In this chapter, we propose to solve this problem by reinforcement learning [12–14]. Note that, to our knowledge, such an approach has never been considered for graph-based semantic image segmentation, although it has been recently studied for graph matching (but in a different context [11,15,16]). Recent related works in computer vision [17] focus on other tasks, such as, for instance, object detection [18], object tracking [19], landmark detection [20], or control of regions of interest in video analysis [20].

This chapter is an extension of a recently proposed approach involving a QAP-based sequential graph matching [21]. The originality and the contribution of this chapter rely on the extension of the experimental results using two datasets and several Deep Neural Networks (DNNs) while studying the influence of a specific parameter of this method, which is the seed size. Section 11.2 describes the proposed method, while Sections 11.3 and 11.4 present experiments and results demonstrating the performance of our approach for the semantic segmentation of 2D and 3D images.

11.2 Reinforcement Learning for Sequential Graph Matching

Figures 11.1 and 11.2 provide an overview of the proposed approach. In the training phase (Figure 11.1), a DNN is trained for image semantic segmentation using an annotated dataset. In the inference phase (Figure 11.2), we propose to correct segmentation errors by using spatial relationships observed between identified regions of the annotated training dataset, leading to an inexact-graph-matching procedure between G_m (built from the training dataset) and G_r (built from the DNN output). When analyzing an unknown image (Figure 11.2), a hypothesis graph G_r is built from the initial DNN

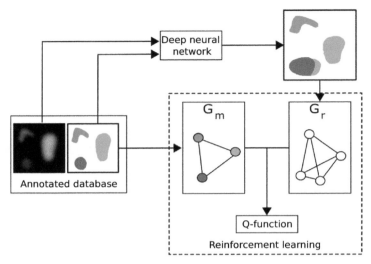

Fig. 11.1. Overview of the training phase, where annotated data are used to train a DNN and learn the model graph. Segmentations over the training data are used with the graph model to learn the Q-function.

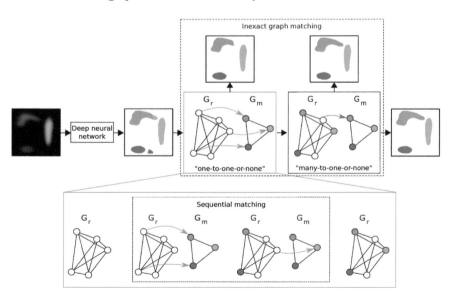

Fig. 11.2. Overview of the inference phase, where segmentation produced by the DNN is used to create image graph. A sequential one-to-one matching is done with a sequential refinement to improve the semantic segmentation.

segmentation result. To identify regions, G_r is matched with G_m, which is an inexact graph-matching problem, as there are more regions in G_r than in G_m due to artifacts. We propose to do this sequentially in two steps. First, an initial "one-to-one" matching is performed to recover one region candidate (vertex of G_r) per class (one vertex of G_m). This is done sequentially according to the ordering learned by reinforcement (based on a Q-function resulting from a preliminary training, Figure 11.1). The second step (refinement), which is not considered in this chapter, focuses on matching remaining artifacts, this being also done sequentially in any order. We hereinafter detail each of these steps.

11.2.1 Graph modeling with semantic segmentation maps

When analyzing an image, the neural network provides a tensor $S \in \mathbb{R}^{P \times N}$ with P the dimensions of the query image (e.g., $P = I \times J$ pixels for 2D images and $P = I \times J \times K$ pixels for 3D images) and N the total number of classes considered for segmentation. At each pixel location p, the value $S(p, n) \in [0, 1]$ is the probability of belonging to class n. The segmentation map $\mathcal{L}^*$ selects the label n of the class with the highest probability:

$$\forall p \in \{1, \ldots, P\}, \quad \mathcal{L}^*(p) = \arg\max_{n \in \{1, \ldots, N\}} S(p, n). \tag{11.1}$$

From $\mathcal{L}^*$, we define a set R of all resulting connected components and finally the graph $G_r = (V_r, E_r, \alpha_v, \alpha_e)$, where V_r is the set of vertices, E_r the set of edges, and α_v and α_e being functions that respectively associate to every vertex and to every edge attribute values depending of the properties considered.

Each vertex $v \in V_r$ is associated with a region $R_v \in R$, with several attributes (hyperparameters of our approach) provided by the function α_v. One of this attributes is key in our method, it is the averaged membership probability vector over the set of pixels $p \in R_v$, therefore computed on the initial tensor S:

$$\forall v \in V_r, \forall n \in \{1, \ldots, N\}, S_{r,v}[n] = \frac{1}{|R_v|} \sum_{p \in R_v} S(p, n). \tag{11.2}$$

The entity $S_{r,v}$ can be assimilated to a probability vector because $\sum_{i \in 1,...,N} S_{r,v}[i] = 1$ and $\forall i \in \{1, \ldots, N\}$, $0 \leq S_{r,v}[i] \leq 1$.

In our method, we are considering complete graphs (i.e., $E_r = V_r \times V_r$), and each edge $e = (i, j) \in E_r$ has an attribute defined by the function α_e (hyperparameter in our method, detailed in the experiments), associated with a relation between the regions R_i and R_j.

The model graph $G_m = (V_m, E_m, \alpha_v, \alpha_e)$ is built from the training set, and V_m is a set of N vertices (one vertex per class) while $E_m = V_m \times V_m$. The structural information that we actually use in our method is described in the experimental section (cf. Sections 11.3.2.2 and 11.4.2.2).

Our method is designed to be generic with regard to the structural information embedded in the graphs G_m and G_r. On the vertices, multiple types of structural information of regions can be used (e.g., maximum distance between two points of a region) in addition to the semantic information that we embed as a vector of dimension N. On the edges, multiple structural relationships between regions (e.g., distance between regions and difference in intensity) can be considered in the graphs. The structural information embedded in the graph G_m is learned by averaging the structural information using the images of the training dataset and its semantic information represented with one-hot encoding vectors.

11.2.2 Sequential one-to-one matching by reinforcement learning

The proposed sequential one-to-one matching between $G_r = (V_r, E_r, \alpha_v, \alpha_e)$ and $G_m = (V_m, E_m, \alpha_v, \alpha_e)$ is formulated as a QAP to be solved sequentially by Q-learning, for finally finding the best assignment X^*:

$$X^* = \arg\min_{X} \{\text{vec}(X)^T K \, \text{vec}(X)\}, \qquad (11.3)$$

where $X \in \{0, 1\}^{|V_r| \times |V_m|}$, X_{ij} means that the vertex $i \in V_r$ is matched to the vertex $j \in V_m$, $\text{vec}(X)$ is the column vector representation of X, and T denotes the transposition operation. The matrix K is defined by

$$K = \lambda\, K_v + (1-\lambda)\, K_e \tag{11.4}$$

and embeds the dissimilarities between the two graphs: K_v embeds the dissimilarities between V_r and V_m (diagonal elements) and K_e embeds the dissimilarities between E_r and E_m (non-diagonal elements). The parameter $\lambda \in [0,1]$ allows weighting the relative contributions of vertex and edge dissimilarities.

For a sequential graph matching, one learns, by reinforcement, from interactions between the agent and the environment [12]. From a given state s_t (set of already matched nodes, at step t of the sequential-matching procedure), the agent (the algorithm) selects and triggers an action (i.e., trying to match a new vertex of $|V_r|$ with a new one of $|V_m|$, or a new subset of vertices). The environment (encompassing image, semantic segmentation, graphs, and graph-matching computations) performs this action and gives back to the agent the resulting new state s_{t+1} (matching result) together with a reward.

In this chapter, the considered reinforcement learning (RL) method is based on Q-learning using a Q-function defined by a Q-Table, that appeared appropriate for preliminary experiments. As underlined in [12], it is widely accepted that such a value-based RL algorithm is appropriate for a discrete RL scenario, which is the case of our graph-matching problem (discrete decision making problem). The design of the agent for our graph-matching problem is detailed hereinafter in terms of state, action, and reward.

11.2.2.1 *State*

As in [11], the state $s_t \in \mathcal{S}$ (at the step t of the episode or matching procedure) is the subset $V_{r,t} \subseteq V_r$ of vertices matched with a subset $V_{m,t} \subseteq V_m$, where $|V_{r,t}| = |V_{m,t}|$, and $\mathcal{S}$ represents all possible partial matchings. The related bijective assignment matrix is $X_t \in \{0,1\}^{|V_{r,t}| \times |V_{m,t}|}$ so that $\forall p \in V_{m,t}, (\sum_{i=1}^{|V_{m,t}|} X_{pi} = 1) \wedge (\sum_{i=1}^{|V_{m,t}|} X_{ip} = 1)$. The matching procedure (episode) goes from $t=0$ ($V_{r,0} = V_{m,0} = \emptyset$) to $t = \infty$ ($|V_{r,\infty}| = |V_m|$ and $V_{m,\infty} = V_m$). We observed experimentally that only a limited number of steps are needed to reach this final situation.

11.2.2.2 Action

The action $a_t \in \mathcal{A}_t$, achieved by the agent at step t, consists in selecting a set of vertices of V_m not in $V_{m,t}$ (i.e., $V_m \setminus V_{m,t}$) and finding the corresponding ones in $V_r \setminus V_{r,t}$. $\mathcal{A}_t$ is the set of possible sets of vertices and depends on t (i.e., already matched vertices at step t are ignored). In our case, at $t = 0$, sets of size larger than one element are considered, while for $t > 0$, single nodes are investigated. The motivation is to begin by finding a small sub-graph matching (seeded graph matching [11]) and then to consider only single nodes to ensure a low complexity. At each step, a QAP optimization is achieved to find the new matching(s), according to Eq. (11.3), where the assignment matrix is initialized according to X_t.

11.2.2.3 Reward

When learning, the agent receives a reward r, based on the quality of the resulting matching. Compared to [11], the reward is not based on the cost related to Eq. (11.3) but on the quality of the resulting semantic segmentation, similar to [20], involving a similarity measurement between the recovered region(s) and the expected one(s). The motivation is to favor the matching with the most similar regions, as several regions (over-segmentation) of the image being analyzed can be associated (and therefore matched) with the same region of the reference segmentation. The reward, depending on both the state s_t and the selected action a_t, is the one considered in [20]:

$$r(a_t, s_t) = \begin{cases} DC + 1 & \text{if } DC > 0.1, \\ -1 & \text{otherwise,} \end{cases} \quad (11.5)$$

where DC is the Dice index between the region(s) associated with the newly matched vertex (or vertices) and the expected one(s).

11.2.2.4 Sequential matching

After the learning procedure leading to the Q function, the matching ordering (i.e., optimal action a_t to be selected at step t) is defined, at each step $t \in [0, \infty]$, by

$$a_t = \arg\max_{a \in \mathcal{A}_t}(Q(s_t, a)), \quad (11.6)$$

where Q ($Q : \mathcal{S} \times \mathcal{A} \to \Re$) is the learned Q-Table [12], representing the maximum expected future rewards for actions at each state ($\mathcal{A}_t \subseteq \mathcal{A}$ in Eq. (11.6)). The Q-function is learned using the same training samples that were used to train the DNN providing the initial segmentation. More precisely, the Q-function is trained by playing a certain number of episodes on the training samples, using the previously defined notions of state, action, and reward. For example, if the training dataset is composed of 5 images to train the DNN and we decide to play 10 episodes per image, the Q-function is trained by playing 50 episodes in order to learn the policy that leads to the one-to-one matching X^I.

11.2.2.5 Complexity

The complexity is directly related to the number of evaluated assignments according to Eq. (11.3), depending on the number of vertices involved and the related set of possible matchings (i.e., set of $X \in \{0,1\}^{|V_r| \times |V_m|}$). Without considering the proposed sequential approach, the number of evaluations NE$_{\text{QAP}}$ equals the following number of $|V_m|$ permutations of $|V_r|$ (without repetitions) or arrangements (i.e., vertex sets from V_r, of size $|V_m|$, to be matched with the V_m vertices):

$$\text{NE}_{\text{QAP}} = P_{|V_m|}^{|V_r|} = \frac{|V_r|!}{(|V_r| - |V_m|)!}. \tag{11.7}$$

With the sequential approach, the number of evaluations NE$_{\text{QAP-RL}}$ is

$$\text{NE}_{\text{QAP-RL}} = P_{|S|}^{|V_r|} + \sum_{i=0}^{|V_m|-|S|} |V_r| - |S| - i, \tag{11.8}$$

where $S \subseteq V_m$ is the set of vertices involved in the first step of graph-matching procedure. Each following step involves only one vertex (right term of Eq. (11.8)). Since $|S| \leq |V_m|$, the number of evaluations can be significantly reduced by minimizing $|S|$ (i.e., $|S| \ll |V_m|$).

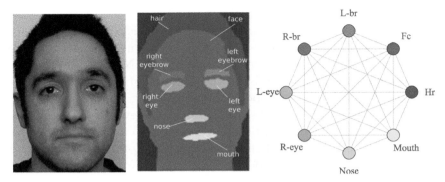

Fig. 11.3. Example of an image from the FASSEG-instances dataset with its annotation and its fully connected graph. The eight classes considered are hair (Hr), face (Fc), left eyebrow (L-br), right eyebrow (R-br), left eye (L-eye), right eye (R-eye), nose, and mouth.

11.3 Semantic Segmentation of 2D Images

In this section, we detail our first experiment, performed on 2D images from the public dataset FASSEG, for the semantic segmentation of face regions such as the eyes, nose, and mouth.

11.3.1 *Dataset*

For this experiment on 2D images, we consider the FASSEG-Instances[b] public dataset that we created for a previous experiment [5]. This dataset is based on the FASSEG[c] public dataset which contains 70 human face images (20 for training and 50 for testing) with associated expert segmentation of face regions (e.g., hair, eyes, nose, and mouth).

The differences between our dataset and the original one come from some modifications we made in order to sub-divide original labels (e.g., right eye and left eye instead of only eyes), leading to eight classes being considered (i.e., hair, face, right eye, left eye, right eyebrow, left eyebrow, nose, and mouth). Figure 11.3 illustrates an image and its annotation from our dataset.

---

[b]https://github.com/Jeremy-Chopin/FASSEG-instances.
[c]FASSEG: https://github.com/massimomauro/FASSEG-repository.

Although the FASSEG dataset includes multiple poses (e.g., frontal or multi-pose), we only consider the frontal images because the considered graph-matching technique may not be robust to face pose changes [5], except if spatial relations are defined in an intrinsic frame and not absolutely, which is out of the scope of our study focusing on QAP optimization. For the sake of simplicity, the term FASSEG is used in the rest of this chapter.

11.3.2 *Evaluation protocol*

For this experiment, we evaluate the influence of the seed size on the results provided by our method in terms of segmentation accuracy and execution time. Moreover, we are also assessing the influence of the seed size on the final results when the DNNs are trained with only a few samples which can, to some extent, lead to more errors in the initial segmentation.

To evaluate the accuracy of the semantic segmentation provided by the DNNs and our approach, we used the Dice score and the Hausdorff distance. The Dice score measures the similarity between the segmented region and the annotated one (value between 0 and 1 to maximize). The Hausdorff distance is a value to be minimized (0 corresponding to a perfect segmentation).

The DNN architectures used for this experiment, the structural information considered in our graphs, and the parameters of the reinforcement learning procedure are detailed in the following sections.

11.3.2.1 *DNNs*

In the context of this experiment, we consider the following DNNs: U-Net [1], U-Net combined with CRF as post-processing [22], PSPNet [23], and EfficientNet [24].

As we are studying the influence of the seed size with DNNs that have been trained with fewer samples, we defined three distinct training configurations (A, B, C). Those training configurations are created using randomly selected data from the initial dataset, and the number of images for each configuration is indicated in Table 11.1. For each configuration, results are averaged over four randomly selected training datasets, leading to four differently trained

Table 11.1. Details of the dataset size for each training configuration (A, B, and C) with the number of samples in the training set, the validation set (in parenthesis), and the testing set.

	Dataset size	A	B	C	Test
FASSEG	70	15 (5)	10 (3)	5 (2)	50

DNN-based neural networks. For instance, for the C configuration on FASSEG, 4 random selections of 5 training images over the 20 available ones are considered. Note that for comparison purpose, the testing set is always the same.

The performance of our approach is evaluated with the Dice score (DSC) and the Hausdorff distance (HD) for each class [25]. The Dice score measures the similarity between the segmented region X and the annotated one Y is defined as

$$\text{DSC}(X, Y) = \frac{2|X \cap Y|}{|X| + |Y|}. \tag{11.9}$$

The Hausdorff distance is a value to be minimized (0 corresponding to a perfect segmentation) and is defined as

$$\begin{aligned} HD(X,Y) &= \max(h(X,Y), h(Y,X)), \\ h(X,Y) &= \max_{x \in X} \min_{y \in Y} ||x - y||. \end{aligned} \tag{11.10}$$

11.3.2.2 Structural information

The class membership probability vector described in Section 11.2 is used as an attribute of the vertices of our graphs. According to this attribute, the dissimilarity function $_vD1_i^k$ ($i \in V_r$ and $k \in V_m$) is used to compute K_v, where

$$_vD1_i^k = \frac{1}{N} \sum_{1 \leq n \leq N} (S_{m,k}[n] - S_{r,i}[n])^2 \tag{11.11}$$

corresponding to the mean squared error (MSE) between the two class membership probability vectors.

For the edges, we are considering structural relationships that are based on both minimum and maximum distances between regions, leading to the following assignment function $D((i,j)) = [d_{\min}^{(i,j)}, d_{\max}^{(i,j)}]$, where

$$d_{\min}^{(i,j)} = \min_{p \in R_i, q \in R_j} (|p - q|), \qquad (11.12)$$

$$d_{\max}^{(i,j)} = \max_{p \in R_i, q \in R_j} (|p - q|). \qquad (11.13)$$

The resulting dissimilarity function using the structural relationships between two edges $(i,j) \in E_m$ and $(k,l) \in E_r$ is defined as follows:

$$_eD_{1(i,j)}^{(k,l)} = \lambda_e \frac{\left(\left|d_{\min}^{(i,j)} - d_{\min}^{(k,l)}\right|\right)}{C_s} + (1 - \lambda_e)\frac{\left(\left|d_{\max}^{(i,j)} - d_{\max}^{(k,l)}\right|\right)}{C_s}, \qquad (11.14)$$

where $\lambda_e \in [0,1]$ is a parameter to balance the influence of the distances and C_s is the maximum diagonal length of the image. For our experiments on the FASSEG dataset, the hyperparameters λ (Eq. (11.4)) and λ_e (Eq. (11.14)) have been arbitrarily set to 0.5.

11.3.2.3 Q-Learning

In order to train the Q-function used in our sequential-matching approach, we considered the same samples as those used to train the DNNs providing the initial segmentation (cf. Table 11.1). We defined 50 episodes of training played for each sample which leads respectively to 750, 500, and 250 episodes played for the training configuration A, B, and C.

To evaluate the influence of the seed size, this experiment focused on a seed S that contains three or two vertices for the first step of the sequential graph matching (the seed composition being learned by reinforcement), while the next steps involve only single vertices.

To evaluate the benefit of learning a specific sequence, we compared our sequential approach where the sequence is learned by reinforcement learning to our sequential approach where the sequence is defined randomly (averaged over 100 random orderings).

11.3.3 *Resilience to small training datasets*

Hereinafter are reported the results of our experiments on the FASSEG dataset. First, we detail quantitative and qualitative results of our approach.

Tables 11.2 and 11.3 report the results of our approach on the FASSEG dataset using a seed composed of three or two vertices, respectively. We observe that the segmentation proposed by our sequential matching improves, to some extent, the quality of the segmentation provided by the DNN, especially on the Hausdorff distance. For example, on average the Dice index and Hausdorff distance for a segmentation provided by a U-Net associated to a CRF using the training configuration B goes respectively from 0.7 and 39.87 to 0.71 and 25.97 (cf. Table 11.2). Moreover, we observe that the results of our sequential-matching method using a sequence learned with reinforcement learning and a sequence defined randomly are, to some extent, similar. However, if we pay attention to the minimal and maximal values in terms of Dice index and Hausdorff distance, we see that learning the sequence provides better results than using a random one, especially when the seed is composed of two vertices (for example, in Table 11.3 for the U-Net associated with a CRF in training configuration B, the minimum Dice index with a learned sequence is 0.56 compared to 0.1 with sequences defined randomly.) Note that results are, to some extent, better when using a seed composed of three vertices compared to a seed only composed of two vertices.

Figure 11.4 illustrates some qualitative results we obtained using our method with a learned sequence (reinforcement) and a random sequence (random ordering) with seeds composed of three or two vertices. These examples qualitatively illustrate, in particular, the relevance of our approach compared to a random ordering. Indeed, for some images, our learned sequence may, to some extent, decrease the quality of the segmentation provided by the DNN (cf. Figure 11.4, middle). However, using a random sequence may lead to an even higher decrease in the segmentation quality (cf. Figure 11.4, top and bottom) as it is reported in Tables 11.2 and 11.3. Nevertheless, learning the sequence is beneficial and provides, to some extent, better results than the random ordering (Figure 11.4, bottom and top images).

Table 11.2. Results on the FASSEG dataset, for different DNNs and training configurations when using a seed composed of three vertices ($S = 3$). Results are given as Dice score (DSC) and Hausdorff distance (HD) with averaged values and standard deviation in parentheses. Note that the minimal and maximal values are also reported in brackets. These values are computed using results from 200 test images (the 50 test images with each DNN trained 4 times) for the learned sequence (reinforcement) and over 20000 test images (the 50 test images with each DNN trained 4 times and using 100 random orderings) for the random sequences (random ordering).

Configuration	A					
Method	DNN		Reinforcement		Random Ordering	
Model	DSC↑	HD↓	DSC↑	HD↓	DSC↑	HD↓
U-Net	0.7 (0.05) [0.55; 0.84]	37.4 (19.32) [13.61; 118.11]	0.7 (0.05) [0.55; 0.84]	26.56 (12.42) [13.97; 87.67]	0.7 (0.05) [0.44; 0.84]	26.65 (12.62) [13.97; 94.87]
U-Net + CRF	0.71 (0.05) [0.57; 0.83]	35.4 (17.18) [12.34; 93.76]	0.71 (0.05) [0.56; 0.83]	24.85 (11.96) [12.34; 77.27]	0.71 (0.05) [0.38; 0.83]	24.82 (11.83) [12.34; 110.06]
PSPNet	0.83 (0.05) [0.64; 0.98]	24.18 (16.58) [2.4; 94.01]	0.83 (0.06) [0.5; 0.98]	19.0 (15.0) [2.4; 118.0]	0.83 (0.07) [0.27; 0.98]	18.88 (14.4) [2.4; 157.48]
EfficientNet	0.84 (0.05) [0.7; 0.98]	22.11 (17.0) [2.38; 82.52]	0.83 (0.07) [0.32; 0.98]	18.15 (15.61) [2.38; 109.16]	0.83 (0.06) [0.21; 0.98]	17.58 (15.07) [2.38; 230.07]

Configuration	B					
Method	DNN		Reinforcement		Random Ordering	
Model	DSC↑	HD↓	DSC↑	HD↓	DSC↑	HD↓
U-Net	0.72 (0.06) [0.58; 0.87]	43.02 (26.64) [12.09; 141.68]	0.73 (0.05) [0.59; 0.87]	25.53 (13.27) [8.66; 81.08]	0.73 (0.06) [0.23; 0.87]	25.63 (13.57) [8.66; 154.86]
U-Net + CRF	0.7 (0.05) [0.56; 0.82]	39.87 (20.71) [14.59; 115.12]	0.71 (0.05) [0.56; 0.83]	25.97 (12.51) [12.46; 79.04]	0.7 (0.05) [0.29; 0.83]	26.09 (12.84) [12.46; 149.28]
PSPNet	0.82 (0.05) [0.66; 0.98]	27.8 (23.0) [2.78; 126.89]	0.82 (0.07) [0.45; 0.98]	19.48 (15.41) [2.78; 137.09]	0.82 (0.07) [0.18; 0.98]	19.43 (15.33) [2.78; 206.1]
EfficientNet	0.83 (0.05) [0.71; 0.98]	26.39 (18.51) [1.6; 81.69]	0.83 (0.06) [0.5; 0.98]	17.93 (13.94) [1.6; 109.39]	0.83 (0.06) [0.18; 0.98]	17.98 (14.63) [1.6; 210.9]

Configuration	C					
Method	DNN		Reinforcement		Random Ordering	
Model	DSC↑	HD↓	DSC↑	HD↓	DSC↑	HD↓
U-Net	0.71 (0.07) [0.48; 0.87]	54.13 (33.5) [6.62; 162.95]	0.72 (0.07) [0.45; 0.87]	29.15 (16.48) [6.12; 96.06]	0.72 (0.07) [0.08; 0.87]	**28.63** (16.9) [6.12; 225.03]
U-Net + CRF	0.69 (0.06) [0.54; 0.83]	60.01 (30.55) [15.65; 157.32]	0.7 (0.06) [0.43; 0.84]	31.33 (15.64) [11.32; 100.25]	0.7 (0.06) [0.11; 0.84]	**31.1** (15.45) [11.32; 223.62]
PSPNet	0.76 (0.09) [0.55; 0.98]	79.93 (47.03) [3.05; 248.97]	**0.77** (0.11) [0.3; 0.98]	**25.76** (18.17) [2.76; 101.23]	0.76 (0.11) [0.12; 0.98]	26.42 (20.64) [2.76; 241.38]
EfficientNet	0.81 (0.05) [0.61; 0.99]	49.8 (33.88) [1.36; 144.91]	0.81 (0.06) [0.49; 0.99]	**19.54** (13.18) [1.36; 107.3]	0.81 (0.07) [0.13; 0.99]	20.51 (17.02) [1.36; 268.06]

Table 11.3. Results on the FASSEG dataset, for different DNNs and training configurations when using a seed composed of two vertices ($S = 2$). Results are given as Dice score (DSC) and Hausdorff distance (HD) with averaged values and standard deviation in parentheses. Note that the minimal and maximal values are also reported in brackets. These values are computed using results from 200 test images (the 50 test images with each DNN trained 4 times) for the learned sequence (reinforcement) and over 20000 test images (the 50 test images with each DNN trained 4 times and using 100 random orderings) for the random sequences (random ordering).

Configuration			A			
Method	DNN		Reinforcement		Random Ordering	
Model	DSC↑	HD↓	DSC↑	HD↓	DSC↑	HD↓
U-Net	0.7 (0.05) [0.55; 0.84]	37.4 (19.32) [13.61; 118.11]	0.7 (0.05) [0.55; 0.84]	26.76 (12.58) [13.97; 87.67]	0.7 (0.05) [0.33; 0.84]	**26.73** (12.84) [13.97; 153.91]
U-Net + CRF	0.71 (0.05) [0.57; 0.83]	35.4 (17.18) [12.34; 93.76]	0.71 (0.05) [0.56; 0.83]	**24.85** (11.96) [12.34; 77.27]	0.71 (0.05) [0.18; 0.83]	24.89 (12.08) [12.34; 207.02]
PSPNet	**0.83** (0.05) [0.64; 0.98]	24.18 (16.58) [2.4; 94.01]	0.82 (0.07) [0.31; 0.98]	20.08 (17.86) [2.4; 157.31]	0.82 (0.07) [0.03; 0.98]	**19.24** (15.87) [2.4; 259.1]
EfficientNet	**0.84** (0.05) [0.7; 0.98]	22.11 (17.0) [2.38; 82.52]	0.83 (0.07) [0.11; 0.98]	18.26 (22.05) [2.38; 278.59]	0.83 (0.06) [0.0; 0.98]	**17.65** (15.96) [2.38; 282.36]

Configuration			B			
Method	DNN		Reinforcement		Random Ordering	
Model	DSC↑	HD↓	DSC↑	HD↓	DSC↑	HD↓
U-Net	0.72 (0.06) [0.58; 0.87]	43.02 (26.64) [12.09; 141.68]	0.73 (0.05) [0.59; 0.87]	**25.52** (13.27) [8.66; 81.08]	0.73 (0.06) [0.23; 0.87]	25.74 (13.79) [8.66; 181.53]
U-Net + CRF	0.7 (0.05) [0.56; 0.82]	39.87 (20.71) [14.59; 115.12]	**0.71** (0.05) [0.56; 0.83]	**26.15** (12.82) [12.46; 79.04]	0.7 (0.05) [0.1; 0.83]	26.31 (13.85) [12.46; 202.49]
PSPNet	0.82 (0.05) [0.66; 0.98]	27.8 (23.0) [2.78; 126.89]	0.82 (0.07) [0.38; 0.98]	19.88 (16.93) [2.78; 128.77]	0.82 (0.07) [0.0; 0.98]	**19.79** (17.31) [2.78; 275.13]
EfficientNet	0.83 (0.05) [0.71; 0.98]	26.39 (18.51) [1.6; 81.69]	0.83 (0.06) [0.47; 0.98]	**17.57** (12.34) [1.6; 65.08]	0.83 (0.06) [0.0; 0.98]	18.35 (16.55) [1.6; 267.88]

Configuration	C					
Method	DNN		Reinforcement		Random Ordering	
Model	DSC↑	HD↓	DSC↑	HD↓	DSC↑	HD↓
U-Net	0.71 (0.07) [0.48; 0.87]	54.13 (33.5) [6.62; 162.95]	0.72 (0.07) [0.48; 0.87]	**28.39** (15.73) [6.12; 96.06]	0.72 (0.07) [0.01; 0.87]	28.64 (16.76) [6.12; 231.8]
U-Net + CRF	0.69 (0.06) [0.54; 0.83]	60.01 (30.55) [15.65; 157.32]	0.7 (0.06) [0.45; 0.84]	**31.43** (16.72) [11.32; 138.08]	0.7 (0.07) [0.0; 0.84]	31.61 (17.06) [11.32; 245.23]
PSPNet	0.76 (0.09) [0.55; 0.98]	79.93 (47.03) [3.05; 248.97]	0.76 (0.1) [0.21; 0.98]	**27.62** (23.9) [2.76; 220.16]	0.76 (0.11) [0.0; 0.98]	27.66 (23.53) [2.76; 285.57]
EfficientNet	0.81 (0.05) [0.61; 0.99]	49.8 (33.88) [1.36; 144.91]	0.8 (0.11) [0.21; 0.99]	23.71 (26.23) [1.36; 172.45]	0.81 (0.08) [0.0; 0.99]	**21.16** (20.86) [1.36; 325.57]

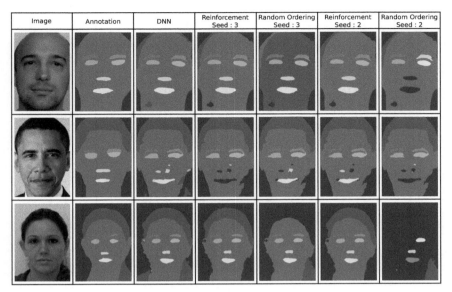

Fig. 11.4. Examples of segmentation results obtained on FASSEG, using our approach with a learned sequence (reinforcement) and a random one (random ordering) with seeds composed of three or two vertices. The annotation and the segmentation of the DNN are also displayed.

11.3.4 *Computation time*

Table 11.4 reports averaged training times (with standard deviation) of the Q-function for each training configuration and DNN when considering a seed composed of three or two vertices. We can observe that the size of the seed has a high influence on the training time of the Q-function. For example, the averaged training time for a Q-function associated with a PSPNet using three vertices for the seed is 45.22s compared to 3.67s when using only two vertices.

Table 11.5 reports the averaged number (with standard deviation) of evaluated assignment matrices ($\overline{\#P}$) (computed according to Eq. (11.8)) and the processing time of an image for our approach when considering a seed S composed of three or two vertices. We can observe that for the training configuration C, which used only few samples to train the DNN, more assignment matrices need to be evaluated compared to other training configurations. For example, the number of assignment matrices to evaluate is on average 1103 for an EfficientNet when using three vertices for the seed in the training

Table 11.4. Averaged training time of the Q-function in seconds with standard deviation for the considered DNNs and for each training configuration.

Method	Training					
Configuration	A		B		C	
Seed	$S=3$	$S=2$	$S=3$	$S=2$	$S=3$	$S=2$
U-Net	28.31 (6.89)	6.45 (0.71)	33.57 (18.19)	5.51 (1.03)	120.61 (183.4)	3.92 (3.25)
U-Net + CRF	37.12 (10.2)	7.71 (1.05)	36.39 (14.85)	5.13 (0.43)	19.45 (9.81)	2.87 (1.1)
PSPNet	**9.19** (1.44)	3.91 (0.68)	9.81 (6.17)	3.58 (0.29)	45.22 (24.03)	3.67 (1.12)
EfficientNet	10.99 (4.06)	**3.78** (1.23)	**6.13** (1.09)	**2.84** (0.69)	**18.23** (13.32)	**2.28** (0.82)

Table 11.5. Averaged number (with standard deviation) of evaluated assignment matrices and measured runtime (in seconds) for each training configuration with a seed composed of three or two vertices.

$S=3$

Configuration	A		B		C	
Method	$\overline{\#P}$	Time	$\overline{\#P}$	Time	$\overline{\#P}$	Time
U-Net	1253.49 (1092.46)	0.33 (0.15)	1884.02 (2060.86)	0.41 (0.26)	**3030.68** (11793.63)	**0.52** (0.83)
U-Net + CRF	1401.27 (1598.48)	0.35 (0.19)	1830.6 (2630.46)	0.39 (0.27)	3341.84 (5265.6)	0.56 (0.46)
PSPNet	1202.88 (2139.09)	0.33 (0.23)	1447.94 (2644.69)	0.36 (0.29)	8329.78 (12049.06)	1.05 (0.96)
EfficientNet	**706.2** (673.11)	**0.26** (0.1)	**1102.98** (1222.38)	**0.33** (0.18)	7144.72 (28088.36)	0.87 (1.95)

$S=2$

Configuration	A		B		C	
Method	$\overline{\#P}$	Time	$\overline{\#P}$	Time	$\overline{\#P}$	Time
U-Net	157.19 (76.58)	0.31 (0.12)	193.55 (120.0)	0.36 (0.18)	**231.5** (255.96)	**0.42** (0.38)
U-Net + CRF	163.33 (97.32)	0.32 (0.15)	186.84 (129.77)	0.35 (0.2)	263.14 (201.74)	0.47 (0.3)
PSPNet	142.86 (108.02)	0.31 (0.18)	155.77 (129.21)	0.32 (0.2)	447.87 (369.37)	0.78 (0.56)
EfficientNet	**111.1** (53.78)	**0.25** (0.09)	**141.15** (84.19)	**0.3** (0.14)	347.56 (497.94)	0.63 (0.75)

configuration B compared to the training configuration C where 7145 matrices need to be evaluated. Moreover, we observe the influence of the seed size on the inference time of our method, which is smaller when considering less vertices for the seed.

11.4 Semantic Segmentation of 3D MRI

In this section, we detail experiments on 3D images using MRI (magnetic resonance imaging) from the IBSR public dataset for the semantic segmentation of sub-cortical regions of the brain.

11.4.1 *Dataset*

The IBSR[d] public dataset is composed of 18 MRIs with the associated expert segmentation of 32 sub-cortical regions of the brain. The limited number of samples available (only 18 3D images) makes this dataset suitable to be used with DNNs able to train on small datasets, such as the U-Net [1]. In this context, we follow the previous work of [26] where only 14 classes (i.e., 14 regions) of the annotated dataset are considered: thalamus (left and right), caudate (left and right), putamen (left and right), pallidum (left and right), hippocampus (left and right), amygdala (left and right), and accumbens (left and right). Figure 11.5 illustrates an image from this dataset and its annotation.

11.4.2 *Evaluation protocol*

As in our previous experiment on the FASSEG dataset, we are assessing the influence of the seed size on the final segmentation accuracy and the execution time of our method with DNNs using only a few samples for training. We are also using the same measures to evaluate the segmentation accuracy of the proposed images, which are the Dice score (DSC) and the Hausdorff distance (HD).

---

[d]The IBSR annotated public dataset can be downloaded at the following address: https://www.nitrc.org/projects/ibsr.

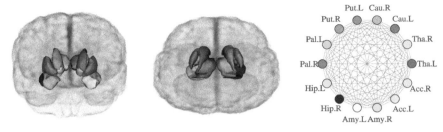

Fig. 11.5. Example of an MRI from the IBSR dataset with its annotation and its fully connected graph. The 3D volumes present the segmentation of brain subcortical regions from two different points of view. Note that not all the sub-cortical regions considered are represented in these images.

Table 11.6. Details of the dataset size for each training configuration (A, B, and C) with the number of samples in the training set, the validation set (in parenthesis), and the testing set.

	Dataset size	A	B	C	Test
IBSR	18	8 (4)	6 (3)	4 (2)	6

The DNN architectures used for this experiment, the structural information considered in our graphs, and the parameters of the reinforcement learning procedure are detailed in the following sections.

11.4.2.1 DNNs

For the IBSR experiment, we considered the following DNNs: U-Net [1] and U-Net combined with CRF as post-processing [22].

As for FASSEG, we considered multiple training configurations (A, B, and C). The numbers of images used to train and test each DNN for each configuration are indicated in Table 11.6. For each configuration, results are averaged over three randomly selected training datasets, leading to three differently trained DNN-based neural networks. For instance, for the C configuration on IBSR, 3 random selections of 4 training images over the 18 available ones are considered. Each of these configurations is however assessed on the same test set for fair comparison (i.e., the same 6 MRIs are used for testing).

The performance of our approach on this experiment is also evaluated with the Dice score (DSC) and the Hausdorff distance (HD) for each class.

11.4.2.2 Structural information

In this experiment, we used as attributes on the vertices of the graphs the class membership probability vectors described in Section 11.2 (cf. Eq. (11.2)) and a region property which is the largest distance between two voxels belonging to the same region, computed as follows:

$$d^i_{\max} = \frac{\max_{(p,q) \in R_i^2}(|p - q|)}{C_s}, \qquad (11.15)$$

where C_s denotes the maximum distance value observed in an image, ensuring that values range within $[0, 1]$.

In this context, to evaluate the dissimilarities between G_r and G_m, and compute K_v, we consider two dissimilarity functions. The first function corresponds to the mean squared error (MSE) between the two class membership probability vectors (cf. Eq. (11.11)). The second dissimilarity function compares the maximal distances within regions as follows:

$$_vD2_i^k = \left| d^k_{\max} - d^i_{\max} \right|. \qquad (11.16)$$

Finally, the matrix K_v is computed by combining the two dissimilarity functions as follows:

$$_vD_i^k = \lambda_v \left(_vD1_i^k \right) + (1 - \lambda_v) \, _vD2_i^k, \qquad (11.17)$$

where $\lambda_v \in [0, 1]$ is a parameter balancing the influence of the class membership probability and the size of the region.

The considered edge attribute is the relative directional position of the centroid of two regions, which is a spatial relationship as in [27]. This relationships is defined as follows:

$$\forall e = (i, j) \in E., \quad \alpha_e(e) = \vec{v}_{ij} = \frac{\overline{R}_j - \overline{R}_i}{C_s}, \qquad (11.18)$$

where E is either E_r (graph G_r) or E_m (graph G_m), and the term $\overline{R}$ denotes the coordinates of the center of mass of region R. This attribute is used to compute the matrix K_e using the following dissimilarity function:

$$eD^{(k,l)}_{(i,j)} = \lambda_e \frac{|\cos\theta - 1|}{2} + (1-\lambda_e)\frac{|\|\vec{v}_{ij}\|_2 - \|\vec{v}_{kl}\|_2|}{C_s}, \quad (11.19)$$

where $\lambda_e \in [0,1]$ is a parameter balancing the influence of the difference in terms of distance and orientation and θ is the angle between the vectors $\vec{v}_{ij}$ and $\vec{v}_{kl}$.

For our experiments on the IBSR dataset, the hyperparameters λ (Eq. (11.4)), λ_v (Eq. (11.17)), and λ_e (Eq. (11.19)) have been arbitrarily set to 0.5.

11.4.2.3 Q-Learning

As in our previous experiment, in order to train the Q-function used in our sequential-matching approach, we considered the same samples as those used to train the DNNs providing the initial segmentation (cf. Table 11.6). We defined 50 episodes of training played for each sample which leads respectively to 400, 300, and 200 episodes played for the training configuration A, B, and C.

To evaluate the influence of the seed size, this experiment focused on a seed S that contains three or two vertices for the first step of the sequential graph matching (the seed composition being learned by reinforcement), while the next steps involve only single vertices.

To evaluate the benefit of learning a specific sequence, we compared our sequential approach where the sequence is learned by reinforcement learning to our sequential approach where the sequence is defined randomly (averaged over 100 random orderings).

11.4.3 Resilience to small training datasets

Tables 11.7 and 11.8 report the results of our approach on the IBSR dataset using a seed composed of three or two vertices, respectively. We can observe that our sequential-matching approach improves the quality of the segmentation provided by the DNN. The improvements are mainly focused on the Hausdorff distance and not significant on the Dice index, which is caused by the fact that classification errors are mainly composed of small regions far from the regions of interest (artifacts). Indeed, our method corrects, to some extent, these small regions by relabeling them, which will have a large influence on the Hausdorff distance, reflecting the important topological correction, but a small influence on the Dice coefficient (due to the small size of the corrected regions). As for the FASSEG experiment, the results of our sequential approach with a sequence learned by reinforcement learning and a sequence defined randomly are similar on average, with only small differences as it can be seen for the training configuration C for the U-Net with a seed composed of three vertices (cf. Table 11.7), where the Hausdorff distance is equal to 5.84 with a learned sequence and 5.78 for the random ones. The main differences are focused on the maximum and minimum values, where our method obtains slightly better results when using a learned sequence than random ones, especially with training configurations composed of only few training samples. For example, in Table 11.8 for a U-Net using the training configuration B, the minimal Dice score observed when using a learned sequence is 0.72 compared to 0.63 when using a random one. Moreover, no significant difference between using a seed composed of three or two vertices has been observed with a sequence learned by reinforcement.

Figure 11.6 illustrates some qualitative results we obtained using our method with a learned sequence (reinforcement) and a random sequence (random ordering), where in this example the random ordering was not able to retrieve a target region (highlighted by the red squares).

11.4.4 Computation time

Table 11.9 reports averaged training times (with standard deviation) of the Q-function for each training configuration and DNN when considering a seed composed of three or two vertices. As for the previous experiment on the FASSEG dataset, using less vertices in the seed leads to a reduction of the training time.

Table 11.10 reports the averaged number (with standard deviation) of evaluated assignment matrices (computed according to Eqs. (11.7) and (11.8)) and the processing time of an image for our approach when considering a seed composed of three or two vertices. As for FASSEG, training configurations composed of few samples such as the configuration C lead to more assignment matrices to be

Table 11.7. Results on the IBSR dataset, for different DNNs and training configurations when using a seed composed of three vertices ($S = 3$). Results are given as Dice score (DSC) and Hausdorff distance (HD) with averaged values and standard deviation in parentheses. Note that the minimal and maximal values are also reported in brackets. These values are computed using results from 18 test images (the 6 test images with each DNN trained 3 times) for the learned sequence (reinforcement) and over 1800 test images (the 6 test images with each DNN trained 3 times and using 100 random orderings) for the random sequences (random ordering).

Configuration						
	A					
Method	DNN		Reinforcement		Random Ordering	
Metrics	DSC↑	HD↓	DSC↑	HD↓	DSC↑	HD↓
U-Net	**0.81** (0.02) [0.76; 0.84]	25.82 (6.93) [14.04; 39.43]	0.8 (0.02) [0.76; 0.83]	4.93 (0.45) [4.03; 5.55]	0.8 (0.02) [0.76; 0.83]	4.93 (0.45) [4.03; 5.55]
U-Net + CRF	**0.81** (0.02) [0.76; 0.84]	25.82 (6.93) [14.04; 39.43]	0.8 (0.02) [0.76; 0.83]	4.93 (0.45) [4.03; 5.55]	0.8 (0.02) [0.76; 0.83]	4.93 (0.45) [4.03; 5.55]

Configuration						
	B					
Method	DNN		Reinforcement		Random Ordering	
Metrics	DSC↑	HD↓	DSC↑	HD↓	DSC↑	HD↓
U-Net	**0.79** (0.03) [0.74; 0.84]	22.99 (7.81) [8.6; 36.91]	0.78 (0.02) [0.72; 0.81]	5.27 (0.46) [4.33; 6.14]	0.78 (0.02) [0.72; 0.81]	5.27 (0.46) [4.33; 6.19]
U-Net + CRF	0.79 (0.02) [0.74; 0.83]	23.49 (7.76) [9.65; 39.29]	0.78 (0.02) [0.74; 0.81]	**5.17** (0.54) [4.23; 6.48]	0.78 (0.02) [0.7; 0.81]	5.18 (0.55) [4.23; 6.86]

(*Continued*)

Table 11.7. (Continued)

Configuration	C					
Method	DNN		Reinforcement		Random Ordering	
Metrics	DSC↑	HD↓	DSC↑	HD↓	DSC↑	HD↓
U-Net	**0.76** (0.03) [0.69; 0.82]	26.94 (8.37) [14.03; 45.38]	0.75 (0.04) [0.69; 0.81]	5.81 (0.79) [4.61; 7.63]	0.75 (0.04) [0.64; 0.81]	**5.78** (0.81) [4.61; 7.63]
U-Net + CRF	0.77 (0.04) [0.7; 0.83]	28.5 (8.43) [13.39; 46.73]	0.76 (0.04) [0.69; 0.82]	5.55 (0.79) [4.34; 7.74]	0.77 (0.03) [0.66; 0.82]	**5.49** (0.77) [4.34; 8.67]

Table 11.8. Results on the IBSR dataset, for different DNNs and training configurations when using a seed composed of two vertices ($S = 2$). Results are given as Dice score (DSC) and Hausdorff distance (HD) with averaged values and standard deviation in parentheses. Note that the minimal and maximal values are also reported in brackets. These values are computed using results from 18 test images (the 6 test images with each DNN trained 3 times) for the learned sequence (reinforcement) and over 1800 test images (the 6 test images with each DNN trained 3 times and using 100 random orderings) for the random sequences (random ordering).

Configuration	A					
Method	DNN		Reinforcement		Random Ordering	
Metrics	DSC↑	HD↓	DSC↑	HD↓	DSC↑	HD↓
U-Net	**0.81** (0.02) [0.76; 0.84]	25.82 (6.93) [14.04; 39.43]	0.8 (0.02) [0.76; 0.83]	4.93 (0.45) [4.03; 5.55]	0.8 (0.02) [0.76; 0.83]	4.93 (0.45) [4.03; 5.55]
U-Net + CRF	**0.81** (0.02) [0.76; 0.84]	25.82 (6.93) [14.04; 39.43]	0.8 (0.02) [0.76; 0.83]	4.93 (0.45) [4.03; 5.55]	0.8 (0.02) [0.76; 0.83]	4.93 (0.45) [4.03; 5.55]

Configuration	B					
Method	DNN		Reinforcement		Random Ordering	
Metrics	DSC↑	HD↓	DSC↑	HD↓	DSC↑	HD↓
U-Net	**0.79** (0.03) [0.74; 0.84]	22.99 (7.81) [8.6; 36.91]	0.78 (0.02) [0.72; 0.81]	5.27 (0.46) [4.33; 6.14]	0.78 (0.02) [0.63; 0.81]	5.27 (0.47) [4.33; 8.68]
U-Net + CRF	**0.79** (0.02) [0.74; 0.83]	23.49 (7.76) [9.65; 39.29]	0.78 (0.02) [0.74; 0.81]	**5.17** (0.54) [4.23; 6.48]	0.78 (0.02) [0.7; 0.81]	5.18 (0.56) [4.23; 6.86]

(*Continued*)

Table 11.8. (Continued)

Configuration				C			
Method	DNN		Reinforcement		Random Ordering		
Metrics	DSC↑	HD↓	DSC↑	HD↓	DSC↑	HD↓
U-Net	0.76 (0.03) [0.69; 0.82]	26.94 (8.37) [14.03; 45.38]	0.75 (0.04) [0.69; 0.81]	5.84 (0.77) [4.61; 7.63]	0.75 (0.04) [0.64; 0.81]	**5.78** (0.8) [4.61; 7.63]
U-Net + CRF	0.77 (0.04) [0.7; 0.83]	28.5 (8.43) [13.39; 46.73]	0.77 (0.04) [0.69; 0.82]	**5.47** (0.82) [4.34; 7.74]	0.77 (0.03) [0.66; 0.82]	5.48 (0.76) [4.34; 8.67]

Table 11.9. Number of evaluated assignment matrices and measured runtime (in seconds in brackets).

Method		Training					
Configuration	A		B		C		
Seed	$S=3$	$S=2$	$S=3$	$S=2$	$S=3$	$S=2$	
U-Net	5461.79 (1691.86)	621.98 (85.6)	7736.51 (6765.74)	858.09 (139.23)	6183.4 (6415.5)	599.59 (81.86)	
U-Net + CRF	6750.77 (2554.15)	614.65 (102.18)	6584.51 (1069.86)	886.26 (75.66)	14176.97 (8041.46)	696.99 (84.05)	

Table 11.10. Number of evaluated assignment matrices and measured runtime (in seconds in brackets).

$S=3$

Configuration	A		B		C	
Method	#P	Time	#P	Time	#P	Time
U-Net	23706.06 (13796.3)	21.93 (6.2)	44171.0 (57329.56)	28.29 (20.96)	31806.16 (24712.77)	25.23 (9.47)
U-Net + CRF	23706.06 (13796.3)	22.01 (6.22)	25572.89 (17503.8)	23.17 (7.17)	61768.78 (47946.28)	35.3 (16.06)

$S=2$

Configuration	A		B		C	
Method	#P	Time	#P	Time	#P	Time
U-Net	1072.56 (377.31)	21.17 (5.6)	1363.78 (1132.03)	25.53 (16.68)	1253.44 (568.79)	23.79 (8.22)
U-Net + CRF	1072.56 (377.31)	21.33 (5.66)	1113.44 (439.95)	22.04 (6.44)	1848.33 (934.21)	32.28 (13.43)

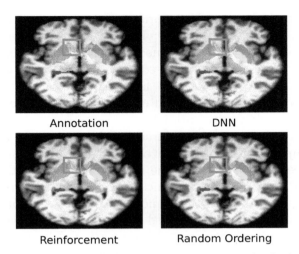

Fig. 11.6. Examples of a slice from a segmentation result obtained on IBSR, using our approach with a learned sequence (reinforcement) and a random one (random ordering). The red squares highlight a target region that the random sequence was not able to retrieve in this case. The annotation and the segmentation provided by the DNN are also displayed.

evaluated compared to training configuration A or B. Reducing the number of vertices in the seed leads to a reduction in the number of matrices to evaluate, which is reflected in the processing time of an image.

11.5 Discussion

Both our experiments show that the use of our QAP-based method improves the semantic segmentation provided by a DNN. These improvements are mainly captured with the Hausdorff distance best suited for capturing segmentation improvements on small areas (cf. Section 11.4). The use of the sequential graph matching allows reducing the computational complexity compared to the traditional QAP (cf. Eqs. (11.7) and (11.8)), depending on the number of vertices considered for the seed.

In those experiments, we compared the results of our method using a sequence learned by reinforcement learning to randomly defined sequences and we observed that on average the results in terms of Dice index and Hausdorff distance are similar. However, the minimal and maximal values obtained, using these metrics, are slightly better when using a learned sequence, which illustrates the importance of the choice of the sequence. From the results in both experiments, using reinforcement learning to learn the sequence is beneficial as it allows defining an order that will assure a coherent topology on the final segmentation.

We studied the influence of the size of the seed on the final results and the computational complexity of our method. As expected, using a smaller seed allows reducing the number of matrices that need to be evaluated (cf. Eq. (11.8)), which is reflected in the reduction of the time required to learn the sequence or process an image. On the FASSEG experiment, using a seed composed of three vertices provides better results than using only two vertices. This is explained by the fact that using more vertices in the seed allows using more structural information at the beginning of the sequential matching thus reducing the possibility of an inappropriate matching (with respect to the structural information available) through the sequential matching. In the IBSR experiment, using a seed composed of three or two vertices does not significantly change the averaged quantitative results we obtained in terms of Dice index and Hausdorff distance, which illustrates that the influence of the seed size is related to the application and to the structural information considered. Finally, as using more vertices in the seed leads to using "more" structural information to avoid incorrect matching, it also leads to an increase in the computational complexity, which suggests to find a compromise that depends on the application and the structural relationships considered.

Our method has however some limitations. Our pipeline is sensitive to change in scale but this type of variation did not occur in our dataset. This problem could be solved by using appropriate structural information that would be resilient to such effect (e.g., relative distance as a function of total brain size instead of absolute spatial distance between brain regions) or considering a deformable graph model [28].

Second, our method is sensitive to missing regions either due to occlusion and varying point of view in the test image or if the DNN fails to accurately perform the preliminary (over) segmentation.

Future directions to address this issue may consist in using multiple model graphs G_m or in using the dueling deep Q-network [29] approach to adapt the matching strategy (i.e., to bypass our limiting requirement for an over-segmentation result from the DNN and introduce revocable actions [11] to correct possible matching mistakes in the earliest stage of the matching) to take advantage of both G_r and information in the segmented test image for a better matching strategy.

Finally, our method is not actually designed for use when multiple instances of the elements described by the graph model occur (e.g., several faces in the image). However, it would be interesting to apply our method to each instance of the element detected in the image, first using an instance segmentation DNN model to isolate each instance from the others.

11.6 Conclusion

In this chapter, we have proposed a reinforcement-learning-based framework for the sequential semantic analysis of image content by exploiting structural information formulated as a QAP-based inexact graph-matching problem.

Our experiments, on the two public datasets FASSEG and IBSR, are promising as they show that our approach dramatically reduces the complexity of this QAP-based inexact graph-matching problem while preserving the efficiency of the analysis. The set of the seed, a specific parameter controlling the number of vertices used at the beginning of the sequential matching, influences the final results and the computational complexity depending on the application and the structural information considered. It would be interesting to study the ability to automatically learn the optimal seed size to be used.

Future works and additional studies will first evaluate our method on other applications with larger datasets and other structural relationships. Another aspect to be studied is the extension of this framework so that the ordering can be dynamically adapted, involving, for instance, the ability to integrate revocable actions [11]. Using a dueling deep Q-network approach [29] would allow adapting the strategy to the current image.

References

[1] T. B. O. Ronneberger and P. Fischer, U-Net: Convolutional networks for biomedical image segmentation, in N. Navab, J. Hornegger, W. Wells and A. Frangi (eds.), *Medical Image Computing and Computer-Assisted Intervention (MICCAI)*. Springer, pp. 234–241 (2015).

[2] I. Bloch, Fuzzy sets for image processing and understanding, *Fuzzy Sets and Systems* **281**, pp. 280–291 (2015), doi:10.1016/j.fss.2015.06.017.

[3] J.-B. Fasquel and N. Delanoue, A graph based image interpretation method using a priori qualitative inclusion and photometric relationships, *IEEE Transactions on Pattern Analysis and Machine Intelligence* **41**(5), pp. 1043–1055 (2019), doi:10.1109/TPAMI.2018.2827939.

[4] K. Kamnitsas, C. Ledig, V. F. J. Newcombe, J. P. Simpson, A. D. Kane, D. K. Menon, D. Rueckert and B. Glocker, Efficient multi-scale 3D CNN with fully connected CRF for accurate brain lesion segmentation, *Medical Image Analysis* **36**, pp. 61–78 (2017), doi:10.1016/j.media.2016.10.004.

[5] J. Chopin, J.-B. Fasquel, H. Mouchère, R. Dahyot and I. Bloch, Semantic image segmentation based on spatial relationships and inexact graph matching, in *2020 Tenth International Conference on Image Processing Theory, Tools and Applications (IPTA)*, pp. 1–6 (2020), doi:10.1109/IPTA50016.2020.9286611.

[6] A. Zanfir and C. Sminchisescu, Deep learning of graph matching, in *2018 IEEE/CVF Conference on Computer Vision and Pattern Recognition*, pp. 2684–2693 (2018), doi:10.1109/CVPR.2018.00284.

[7] D. Bacciu, F. Errica, A. Micheli and M. Podda, A gentle introduction to deep learning for graphs, *Neural Networks* **129**, pp. 203–221 (2020), doi:10.1016/j.neunet.2020.06.006.

[8] L. Ziyao, Z. Liang and S. Guojie, Gcn-lase: Towards adequately incorporating link attributes in graph convolutional networks, in *28th International Joint Conference on Artificial Intelligence (IJCAI)*, pp. 2959–2965 (2019), doi:10.24963/ijcai.2019/410.

[9] J.-B. Fasquel and N. Delanoue, Approach for sequential image interpretation using a priori binary perceptual topological and photometric knowledge and k-means-based segmentation, *Journal of the Optical Society of America A* **35**(6), pp. 936–945 (2018), doi:10.1364/JOSAA.35.000936.

[10] G. Fouquier, J. Atif and I. Bloch, Sequential model-based segmentation and recognition of image structures driven by visual features and

spatial relations, *Computer Vision and Image Understanding* **116**(1), pp. 146–165 (2012), doi:10.1016/j.cviu.2011.09.004.

[11] C. Liu, R. Wang, Z. Jiang, J. Yan, L. Huang and P. Lu, Revocable deep reinforcement learning with affinity regularization for outlier-robust graph matching (2020), *CoRR*, https://arxiv.org/abs/2012.08950.

[12] R. S. Sutton and A. G. Barto, *Reinforcement Learning: An Introduction*, 2nd edn. The MIT Press, Cambridge (2018), http://incompleteideas.net/book/the-book-2nd.html.

[13] Y. Yang and A. Whinston, A survey on reinforcement learning for combinatorial optimization (2020), arXiv:2008.12248 [cs.LG], https://arxiv.org/abs/2008.12248.

[14] D. Yan, J. Weng, S. Huang, C. Li, Y. Zhou, H. Su and J. Zhu, Deep reinforcement learning with credit assignment for combinatorial optimization, *Pattern Recognition* **124**, p. 108466 (2022), https://doi.org/10.1016/j.patcog.2021.108466, https://www.sciencedirect.com/science/article/pii/S0031320321006427.

[15] B. Wu and L. Li, Solving maximum weighted matching on large graphs with deep reinforcement learning, *Information Sciences* **614**, pp. 400–415 (2022), https://doi.org/10.1016/j.ins.2022.10.021, https://www.sciencedirect.com/science/article/pii/S0020025522011410.

[16] H. Wang, Y. Zhang, L. Qin, W. Wang, W. Zhang and X. Lin, Reinforcement learning based query vertex ordering model for subgraph matching, in *2022 IEEE 38th International Conference on Data Engineering (ICDE)*, pp. 245–258 (2022), doi:10.1109/ICDE53745.2022.00023.

[17] N. Le, V. S. Rathour, K. Yamazaki, K. Luu and M. Savvides, Deep reinforcement learning in computer vision: a comprehensive survey, in *Artificial Intelligence Review* **55**, p. 2733–2819 (2022), doi:10.1007/s10462-021-10061-9.

[18] A. Pirinen and C. Sminchisescu, Deep reinforcement learning of region proposal networks for object detection, in *2018 IEEE/CVF Conference on Computer Vision and Pattern Recognition*, pp. 6945–6954 (2018), doi:10.1109/CVPR.2018.00726.

[19] S. Yun, J. Choi, Y. Yoo, K. Yun and J. Y. Choi, Action-driven visual object tracking with deep reinforcement learning, *IEEE Transactions on Neural Networks and Learning Systems* **29**(6), pp. 2239–2252 (2018), doi:10.1109/TNNLS.2018.2801826.

[20] M. Sun, J. Xiao, E. G. Lim, Y. Xie and J. Feng, Adaptive ROI generation for video object segmentation using reinforcement learning, *Pattern Recognition* **106**, p. 107465 (2020), doi:10.1016/j.patcog.2020.107465.

[21] J. Chopin, J.-B. Fasquel, H. Mouchère, R. Dahyot and I. Bloch, QAP optimisation with reinforcement learning for faster graph matching in sequential semantic image analysis, in *International Conference on Pattern Recognition and Artificial Intelligence (ICPRAI)*, pp. 47–58 (2022), doi:10.1007/978-3-031-09037-0_5.

[22] S. Zheng, S. Jayasumana, B. Romera-Paredes, V. Vineet, Z. Su, D. Du, C. Huang and P. H. S. Torr, Conditional random fields as recurrent neural networks, in *2015 IEEE International Conference on Computer Vision (ICCV)*, pp. 1529–1537 (2015), doi: 10.1109/ICCV.2015.179.

[23] H. Zhao, J. Shi, X. Qi, X. Wang and J. Jia, Pyramid scene parsing network, in *2017 IEEE Conference on Computer Vision and Pattern Recognition (CVPR)*, pp. 6230–6239 (2017), doi:10.1109/CVPR.2017.660.

[24] M. Tan and Q. Le, EfficientNet: Rethinking model scaling for convolutional neural networks, in K. Chaudhuri and R. Salakhutdinov (eds.), *Proceedings of the 36th International Conference on Machine Learning*, Vol. 97. Proceedings of Machine Learning Research, pp. 6105–6114 (2019).

[25] K. Kushibar, S. Valverde, S. González-Villà, J. Bernal, M. Cabezas, A. Oliver and X. Lladó, Automated sub-cortical brain structure segmentation combining spatial and deep convolutional features, *Medical Image Analysis* **48**, pp. 177–186 (2018), https://doi.org/10.1016/j.media.2018.06.006, https://www.sciencedirect.com/science/article/pii/S1361841518303839.

[26] K. Kushibar, S. Valverde, S. González-Villà, J. Bernal, M. Cabezas, A. Oliver and X. Lladó, Automated sub-cortical brain structure segmentation combining spatial and deep convolutional features, *Medical Image Analysis* **48**, pp. 177–186 (2018), doi:10.1016/j.media.2018.06.006.

[27] A. Noma, A. B. Graciano, R. M. Cesar Jr, L. A. Consularo and I. Bloch, Interactive image segmentation by matching attributed relational graphs, *Pattern Recognition* **45**(3), pp. 1159–1179 (2012), doi: 10.1016/j.patcog.2011.08.017.

[28] F. Zhou and F. De la Torre, Deformable graph matching, in *Proceedings of the IEEE Conference on Computer Vision and Pattern Recognition (CVPR)* (2013).

[29] Z. Wang, T. Schaul, M. Hessel, H. Van Hasselt, M. Lanctot and N. De Freitas, Dueling network architectures for deep reinforcement learning, in *33rd International Conference on International Conference on Machine Learning*, Vol. 48, p. 1995–2003 (2016).

Chapter 12

FEM and Multi-Layered FEM: Feature Explanation Methods with Statistical Filtering of Important Features

Alexey Zhukov, Jenny Benois-Pineau*, Romain Giot,
Romain Bourqui, and Luca Bourroux

*University of Bordeaux, CNRS, Bordeaux INP, LaBRI, UMR5800,
33400 Talence, France*
*jenny.benois-pineau@u-bordeaux.fr

Abstract

Deep learning (DL) approaches have become essential in data analysis and classification but they appear as black boxes, the results being given without any explanation. Then, the need for explanations of Deep Neural Network (DNN) decisions has led to an active research in eXplainable Artificial Intelligence (XAI). Here we developed Multi-Layered FEM that is an improvement of the Feature Explanation method that was based on the activation values of the target layer of the network.

12.1 Introduction

Deep learning (DL) approaches have become essential in data analysis and classification for various applications including healthcare [1] and natural language processing [2]. While DL models have produced impressive results, they lack transparency, which has prevented their widespread use in critical applications, such as medical image analysis or security.

The need for explanations of Deep Neural Network (DNN) decisions has led to an active research in eXplainable Artificial Intelligence (XAI) [3]. In the field of pattern recognition for images and videos, one way of explaining a DNN's decision consists of identifying the set of input pixels that have contributed the most to the decision [4]. Building such an explanation and comparing it to what is expected by a human being helps increase human trust in the classifier. For example, Ribeiro et al. [5] showed how a trained classifier wrongly used the presence of snow as the distinguishing feature between the "Wolf" and "Husky" classes, which could have been identified through an explanation of the classifier's decision. To address such issues, we previously proposed Feature Explanation Method (FEM)[a] that computes such explanation using the activation values of a target layer. On its basis, we developed Multi-Layered FEM [6] which combines several FEM explanations from different layers of the classifier network to increase the resolution of the explanation map. Both methods are gradient-free and are based on the selection of strong features in convolutional layers. They use a strong hypothesis of underlying Gaussian distribution of features in each feature channel in a feature tensor resulting from a convolutional layer, which cannot be always held.

This chapter presents and extends our previous works (detailed in Sections 12.3 and 12.5) by analyzing more deeply a key component of FEM and MLFEM: the feature filtering strategy. Our new contributions are the following:

(i) the study and verification of Gaussian distribution hypothesis of features in FEM and MLFEM,
(ii) the proposal of a new strategy for filtering out weak features in both methods. The latter uses distributions of known shape and computes the filtering threshold on the theoretical cumulative distribution function (cdf). In case when the hypothesis of any known distribution is not held, the threshold is computed on experimental probability density function (pdf).

---

[a]Available at https://github.com/labribkb/fem/blob/main/FEM.ipynb.

This chapter is organized as follows. Section 12.2 summarizes the state of the art in explanation methods. Section 12.3 reminds the principle of the Feature Explanation Method (FEM). Section 12.4 describes the selection of important features with statistically adaptive thresholds. Section 12.5 presents the MLFEM. Section 12.6 gives the evaluation protocol, while results are reported in Section 12.7. Conclusion and perspectives are drawn in Section 12.8.

12.2 State of the Art

The methods used to explain decisions of DNNs rely on the extraction of information from the most important elements in the layers of deep neural networks in a variety of ways. They often render the results as a heatmap overlaying the input data, e.g., images, explaining important regions in the input to the user. Such heatmaps are called "saliency maps" [7]. Nevertheless, this term sometimes is misleading, as it may designate visual saliency maps obtained from psycho-visual experiments with human observers. Thus, in the following, we also use the term "explanation map."

Ayyar et al. [4] analyzed a rich set of explanation methods, highlighting important patterns from the input image. Following their proposed taxonomy, we can identify "black-box" and "white-box" families of explanation methods. The "black-box" methods remain model agnostic and are applicable to any classifier, as they identify important pixels in images by masking different parts of the input and tracking induced decisions (e.g., LIME [5] method or H^2O [8]). "White-box" methods, on the contrary, use the internal architecture of DNNs. From now on, we focus on this family.

One of the earliest of these methods was the so-called Deconvolution Network (DeconvNet) proposed by Zeiler et al. [9]. The principle here was to build the mapping of the output score to the input space using reverse filters or "deconvolution," thus identifying the important pixels.

The methods based on gradient back-propagation such as the popular GradCam [7], or its further improved versions (e.g., SmoothGrad [10] or integrated gradients [11]), operate by propagating gradients from the network's output to the last convolutional layer. The important input information is located where the gradients are strong.

The *Layered Relevance Propagation* (LRP) [12] method is also based on the same idea of back propagation but without the need of gradient computations. Here, the relevance of neurons from the last decision layer is propagated through receptive fields in previous layers using the principle of conservation of the relevance at each layer which allows identifying important input neuron pixels. Dedicated rules have to be written for each family of network layers.

Feature Explanation Method (FEM) [13] (for more details, see Section 12.3) is also based on backpropagation, but it is not linear in the sense that with the help of statistical filtering of the features of the last convolutional layer it identifies the most important ones. FEM selects the most import activation values by using the $k-\sigma$ rule that assumes a Gaussian distribution of the features in each feature map of the feature tensor from the last convolutional layer. In reality, activation distribution might not be Gaussian, so it is needed to verify if this shortcut impacts the quality of the explanation.

Propagating information through subsequent layers, the Deep NNs lose high resolution information due to the cascaded convolutions and sub-sampling. GradCam-like methods as well as FEM-like methods suffer from this issue as they use information from the latest layers of the network. LRP-like methods or deconvolution-like ones do not suffer from it as they propagate the information up to the input layer. Hence, it makes sense to explore the DNN classifier a bit more, preserving important details in each convolutional layer which finally brings the classification decision. We assert that the fusion of explanations from several layers is a key to improve the final explanation. For this reason, we proposed an extension of FEM, the Multi-Layered FEM (MLFEM) [6], that relies on information fusion on feature importance from different layers.

12.3 Feature Explanation Method (FEM)

Feature Explanation Method (FEM) [13] is a recent algorithm used to produce an explanation map of the decision of a CNN. Opposite to other methods from the literature, it is class agnostic and does not need to provide a class of interest. FEM makes two hypotheses: (i) Strong features at the last convolutional layer will contribute the most in the final decision of the CNN in a classification task when pushed through fully connected layers. (ii) It assumes that the features in each feature map in a feature tensor follow a Gaussian distribution; this is a simplification hypothesis that could hold in case of large feature maps. At the last layer of convolution, the strongest features are the representations of the most relevant regions in the input image. This comes from the interpretation of the "convolutional" part of a deep CNN as of a multi-scale pyramid with filtering, nonlinear input signal transformations and sub-sampling [14].

By analyzing only the last layer of the CNN part of the model, one can get input pixels which contributed the most into the final decision. FEM uses activations of the last layer of the CNN *after* the nonlinearity (most likely *ReLu*). Thus, only *positive* features will be picked up. As FEM is class agnostic, it is not a problem if it emphasizes on some high activations that are negatively weighted in the next layer. Indeed, the features remain important for the overall classification whether they vote for or against a specific class.

The last convolutional layer produces activations of size $(W \times H)$ for D feature maps f_i, $i = 1 \ldots D$. D binary maps, b_i, corresponding to each of the D feature maps f_i are computed by selecting the strongest features of f_i:

$$b_i(x,y) = \begin{cases} 1 & \text{if } f_i(x,y) \geq \mu_i + k\sigma_i, \\ 0 & \text{otherwise.} \end{cases} \quad (12.1)$$

Mean μ_i and standard deviation σ_i are individually estimated for each feature map f_i in the feature tensor. Equation (12.1) is based on underlying hypothesis of Gaussian distribution of features in each feature map f_i. The coefficient k is regulated by the *k-sigmas* rule and chosen as $k = 2$, thus considering as important, 0.025% of features constituting the right distribution cue.

The contribution of each map into the decision is also weighted by a map-importance weight corresponding to μ_i. The resulting feature importance map s is computed as a linear combination of binary channel maps with corresponding weights:

$$s(x,y) = \sum_{i=1}^{D} b_i(x,y)\mu_i. \quad (12.2)$$

It has the dimension of $(W \times H)$ of the feature maps and has to be upscaled to the original resolution of the input image by linear interpolation. A min-max normalization is finally used to bring the domain from $\mathbb{R}^+$ to $[0, 1]$ and obtain the final normalized map of feature importance S:

$$S = \text{normalization}\,(\text{upscale}\,(s)). \tag{12.3}$$

12.4 Selection of Important Features with Statistically Adaptive Thresholds

Many activation values are zero in each feature map in the feature tensor generated by the *ReLu* function. Thus, the Probability Density Function (pdf) of the features in each channel systematically has a strong pick for zero value (see the red histogram in Figure 12.1). Due to this, the Gaussian distribution of the features does not hold in practice. Although this assumption allows FEM to have good results, it is interesting to check if other strategies of threshold selection that better fit the distribution of activation values could improve the quality of the generated explanation maps. One way to do this is to test statistical hypotheses of the fit of the experimental pdf to parametric

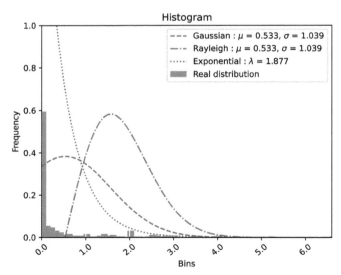

Fig. 12.1. Example of a distribution of features, on one of the channels of some model layer, with superimposed parametric distributions.

distributions which contain a pronounced pick close to zero. In the case when the sample distribution fits to one of these distributions, the choice of the threshold at a given probability level (2.5% in our case) might be more precise. Hence, we select exponential [15] and Rayleigh [15] laws and also conduct statistical hypothesis test for Gaussian distribution [15]. If the features in the feature map follow a distribution of a known shape with the parameters estimated on the data sample, we can use the corresponding parameterized law to compute the feature threshold for binarization. It is done according to the known Cumulative Distribution Function (cdf). Otherwise, we propose to compute the threshold on the experimental pdf of the given feature map. In the follow-up of this section, we describe the statistical hypothesis tests we conduct and propose an adaptive threshold to compute on the experimental law in case when the hypothesis of known distribution has to be rejected.

12.4.1 *Approximation of experimental feature distribution laws with selection of PDFs*

We consider three probability density functions (pdfs) for our statistical tests: Gaussian, Rayleigh, and exponential. We chose these laws, as their pdfs have or can have a pronounced pick close to zero, which is the case of feature histograms in feature maps.

Gaussian density function:

$$F_g(x; \mu_g, \sigma_g) = \frac{1}{\sqrt{2\pi\sigma_g^2}} \exp\left(-\frac{(x-\mu_g)^2}{2\sigma_g^2}\right), \quad (12.4)$$

where x is a random variable and μ_g and σ_g are the mean and standard deviation of activation values. We estimate them on the sample and further denote as $\hat{\mu}_g, \hat{\sigma}_g$.

Rayleigh density function:

$$F_r(x; \mu_r, \sigma_r) = \frac{x-\mu_r}{\sigma_r^2} \exp\left(-\frac{(x-\mu_r)^2}{2\sigma_r^2}\right), \quad (12.5)$$

where x is a random variable and μ_r and σ_r are estimated on the sample as stated above ($\hat{\mu}_r, \hat{\sigma}_r$).

Exponential density function:

$$F_e(x; \lambda_e) = \begin{cases} \lambda_e \exp(-\lambda_e x) & \text{if } x \geq 0, \\ 0 & \text{otherwise,} \end{cases} \qquad (12.6)$$

where λ_e is the rate parameter. In our work, x is a non-negative variable, and thus only the first line on the right side of the equation is considered. The rate parameter λ_e is defined as $\lambda = \frac{1}{\mu_e}$. We also estimate μ_e on the sample and further denote it as $\hat{\mu}_e$.

As an example, we present in Figure 12.1 the histogram of features in an arbitrary chosen feature map (the 64th feature channel map of 8th conv block) of ResNet50 when classifying an image from Salicon dataset [16]. The hypothesized distributions: Gaussian, Rayleigh, and exponential with parameters estimated on this feature map are plotted as well.

12.4.2 Test of statistical hypotheses of underlying distribution

To select the underlying distribution of known shape, i.e., Gaussian, Rayleigh, or exponential, we conduct statistical tests. Among different statistical tests, we use the Kolmogorov–Smirnov (K–S) test [17] of distribution fitness as it is quick to compute and proved to be efficient in various applications.

It is a non-parametric statistical test used to determine whether a given sample of data comes from a particular probability distribution. The null hypothesis of the K–S test is that the sample follows the hypothesized distribution, while the alternative hypothesis is that it does not. The test compares the cdf of experimental law, $\hat{\Phi}$, where $\hat{\Phi}(t)$ is the sum from 0 to t of the histogram, with the cdf of theoretical law, Φ, which is assumed to be known.

The K–S test statistic, denoted by $\mathfrak{K}$, is defined as the maximum absolute difference between the empirical cumulative distribution function and the theoretical cdf as

$$\mathfrak{K} = \max_x |\hat{\Phi}(x) - \Phi(x)|, \quad x \in \mathbb{R}. \qquad (12.7)$$

If $\mathfrak{K} > \mathfrak{K}_{\text{crit}}$, where $\mathfrak{K}_{\text{crit}}$ is the critical value for the specified significance level α, then the null hypothesis is rejected. The critical value $\mathfrak{K}_{\text{crit}}$ is the value of Kolmogorov distribution for the given significance level α, typically 0.95. We refer the reader to [17] for the detailed description. Otherwise, the p-value is computed which is the probability of obtaining a value of $\mathfrak{K}$ at least as extreme as the one observed, assuming that the null hypothesis is true. If the p-value is less than a specified threshold, typically 0.05, then the null hypothesis is rejected, and it is concluded that the sample does not follow the hypothesized distribution.

12.4.3 Data quantization and simplification for statistical tests

To perform the test, we have to quantize our data to build corresponding pdf and cdf, respectively. These data correspond to the activation values of one channel of one convolutional layer for one image, otherwise we call it "feature map." The process is individually repeated for each channel. Furthermore, in some channels, the features are of a very small value, which means that they should not contribute to the final decision by the classification layer of the CNN classifier. Hence, in the follow-up, we discuss feature quantization and drop-out of feature maps of low magnitude.

We have chosen *linear quantization* to quantize features in the feature maps. It is defined as follows:

$$x_q = \left(\left\lfloor \frac{x}{q} \right\rfloor + 0.5 \right) \cdot q, \tag{12.8}$$

where x is a continuous feature value and q is the quantization step. The output of this function is a discrete value that represents the bin in which the continuous value falls. In this study, based on the analysis of the feature magnitude in different feature maps of our dataset, we have selected the quantization step as $q = 0.1$. According to our experiments, finer quantization steps do not bring improvement in hypothesis testing. Stronger values of quantization step q make the distribution very rough, as the feature values are generally small.

Data simplification consists in dropping out feature maps which contain only zero values after quantization. These feature maps do not contribute into the explanation map anyway. When discarding zero-valued feature maps, we save computational overload for statistical tests.

A feature map f_i, $i = 1 \ldots D$, is discarded if the all its values equal zero. It will not be used in the remaining part of our approach.

The remaining feature channels are used to estimate the cdf, which is compared to the cdf of the theoretical distribution using the Kolmogorov–Smirnov (K–S) test.

If the null hypothesis of the K–S test is rejected for all cdfs we consider, this means that the underlying feature distribution is far from the theoretical law. Hence, in this case, we propose an adaptive threshold to compute on the experimental law.

12.4.4 Threshold computation with theoretical or experimental distributions

Our method of threshold choice for selection of important features in each feature map can be resumed in the following steps:

(1) Quantize the features in the map f_i.
(2) Drop out the feature map f_i if all it values equal zero.
(3) Otherwise, compute the histogram h_i.
(4) Estimate mean $\hat{\mu}_i$ and standard deviation $\hat{\sigma}_i$ and calculate cdfs of three distributions:

- Gaussian cdf

$$\Phi_g(x; \hat{\mu}_i, \hat{\sigma}_i) = \frac{1}{2}\left[1 + \text{erf}\left(\frac{x - \hat{\mu}_i}{\hat{\sigma}_i \sqrt{2}}\right)\right], \tag{12.9}$$

with

$$\text{erf}(y) = \frac{2}{\sqrt{\pi}} \int_0^y \exp\left(-t^2\right) dt. \tag{12.10}$$

- Rayleigh cdf

$$\Phi_r(x; \hat{\mu}_i, \hat{\sigma}_i) = 1 - \exp\left(-\frac{(x - \hat{\mu}_i)^2}{2\hat{\sigma}_i^2}\right). \tag{12.11}$$

- Exponential cdf

$$\Phi_e(x; \lambda_i) = \begin{cases} 1 - \exp\left(-\lambda_i x\right) & x \geq 0, \\ 0 & x < 0. \end{cases} \tag{12.12}$$

(5) Execute K–S tests for the three theoretical cdfs.
(6) If p-value of the test is higher than a given confidence level ($\alpha = 0.05$) for one of the theoretical distributions, then compute the threshold th according to the theoretical cdf of the corresponding distribution and $\epsilon = 0.025$:

- Gaussian th

$$\text{th}_g(\epsilon; \hat{\mu}_i, \hat{\sigma}_i) = \hat{\mu}_i + \hat{\sigma}_i * \sqrt{2}\text{erf}^{-1}(2\epsilon - 1), \quad (12.13)$$

with

$$\text{erf}^{-1}(y) = \sum_{k=0}^{\infty} \frac{c_k}{2k+1} \left(\frac{\sqrt{\pi}}{2}\right)^{2k+1} \quad (12.14)$$

and

$$c_0 = 1; c_k = \sum_{m=0}^{k-1} \frac{c_m c_{k-1-m}}{(m+1)(2m+1)}. \quad (12.15)$$

- Rayleigh th

$$\text{th}_r(\epsilon; \hat{\mu}_i, \hat{\sigma}_i) = \hat{\mu}_i + \hat{\sigma}_i * \sqrt{-2\ln(1-\epsilon)}. \quad (12.16)$$

- Exponential th

$$\text{th}_e(\epsilon; \lambda_i) = \frac{-\ln(1-\epsilon)}{\lambda_i}. \quad (12.17)$$

- Otherwise compute the th on experimental pdf h_i as

$$\text{th}(\epsilon) = t(\epsilon) \text{ such that } \sum_{t=0}^{t=t_\epsilon} h_i(t) \leq 1 - \epsilon, \quad (12.18)$$

with t the bin index in the histogram h_i of ith feature map.

In Section 12.7, we report on our experiments on the threshold estimation on the used datasets Salicon [16] and Cat2000 [18]. In the follow-up, we introduce the multi-layered FEM (MLFEM). It uses FEM in a layer-wise manner inside the DNN and combines explanation maps of conv layers in the resulting map, thus integrating the importance of features in all considered convolutional layers. We also discuss the selection of the threshold on feature maps for this method.

12.5 Multi-Layered Feature Explanation Method

Our method Multi-Layered Feature Explanation Method (MLFEM) [6] is built upon FEM [13]. The latter relies on the analysis of activations at the *last* conv layer of a CNN classifier. As each layer of a CNN embeds information at a different scale, we assume that computing FEM at several layers and fusing them would improve the quality of the feature attribution; this is the main idea of MLFEM. In the following, we briefly review FEM and describe the adaptations for MLFEM.

12.5.1 *Principles of multi-layered FEM (MLFEM)*

FEM as presented previously in Section 12.3 can be applied on any layer of a CNN. We can merely pretend to truncate the model at a particular layer and see that FEM would work as is. The application of FEM on a CNN consisting of L convolutional layers will yield L different feature importance maps. As all importance maps are interpolated in FEM, we finally have L maps of the input resolution. The information provided by the maps is layer-dependent, and it is interesting to fuse them. Now the question is how to obtain a single heatmap for the input, highlighting the pixels which have contributed into the network decision the most.

Let M be our convolutional neural network, with $l = 1, \ldots, L$ convolutional layers. Let us denote F_l a feature tensor obtained at each convolutional layer after the positive nonlinearity ($ReLu$).

Let us denote by $H(F_l)$ the operator which implements FEM yielding to a normalized importance map S_l of features F_l. The multi-layered FEM pixel importance map is obtained by fusing all the importance maps S_l with a fusion operator $\bigoplus$: $S = \bigoplus_{l=1}^{L} S_l$.

The combination of L different maps can be done in a recursive manner. For each intermediate map, S_l, we construct S as the combination of S and the current intermediate map S_l. We then move along for each $l < L$.

An alternative is to devise a fusion operator that takes a variable number of arguments to produce a single final feature importance map.

The reasoning for applying multiple times the same explanation method at different network's layers is the following. The network

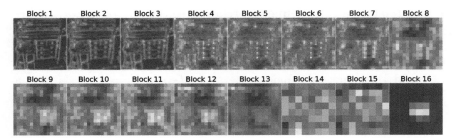

Fig. 12.2. FEM applied on every convolutional block of a typical ResNet50 architecture. Resolution is higher for the first layers. Salicon dataset.

passes the input image through multiple layers of convolution. The convolution layers produce results that are position invariant. They are meant to pick up on spatially local feature. With each step deeper in the network, the convolutional layer picks up on more and more abstract concepts (see Figure 12.2). The very first layers are generally performing edge detection while the later ones extract abstract concept like "face," "car," etc.

Different information is available at different points in the network; by combining the different activation maps at different points in the network, we can reconstruct a heatmap that takes advantage of all this scattered information.

12.5.2 *Fusion operators*

Quite a number of fusion operators are used in data fusion in classification tasks [19]. In our present work, we have appealed to the algebraic fusion operators and to the fusion by a convolutional neural network trained with regard to the ground truth obtained from human observers of the content.

12.5.2.1 *Algebraic fusion operators*

Algebraic fusion operators applied in our work are presented in the following. They are applied element-wise on feature maps and correspond to the following:

- The *max* operator $max(a,b)$ is the result of the *max* operation applied element-wise to a and b. $max_{u,v}(a,b) = max(a_{u,v}, b_{u,v})$.
- The weighted addition $add(a,b)$ is the result of the addition of a and b given a factor α. $add_{u,v}(a,b) = \alpha \cdot a_{u,v} + (1-\alpha) \cdot b_{u,v}$.

- The *top* operator is defined in relation to the *add* operator, taking only the highest features of b. $b'_i = b_i$ if $(b_i > \mu(b))$, 0, otherwise, top(a, b) = add(a, b′).
- In the *fem* operator, we can produce the same result as the FEM would by using this fusion operator. $fem(a, b) = b$; it is also a special case of the *add* operator with $\alpha = 0$.

The maximum is commutative: $max(a, b) = max(b, a)$, but the *add* and the *top* operators are not: $add(a, b) \neq add(b, a)$, $top(a, b) \neq top(b, a)$.

The *add* operation in fact constructs a geometric sequence. $\Sigma \alpha (1-\alpha)^{l-1} S_l$. The normalization operator can also be applied, either at the end of the fusions or interleaved between each binary operator.

Due to the fact that *add* and *top* are not commutative, they take advantage of the structure of data, namely of the fact that the first operand of the operator is the cumulated map in the recursive fusion approach and can be more or less taken into account regulated by the parameter α.

12.5.2.2 *Fusion by a convolutional neural network*

The idea here consists in training a light CNN m whose input is the set of feature importance maps S_l from all layers of the CNN model M to be explained, interpolated to the resolution of input images of M. The training of m is fulfilled with regard to the ground truth expressing human perception of the visual content in the classification task. This perception is measured by Gaze Fixation Density Maps (GFDMs) obtained in a psycho-visual experiment when human subjects observe images to classify them. We refer the reader to [20] for a detailed explanation of such an experiment. Human gaze fixations from a number of observers are recorded for each image by an eye-tracking device.

Then on each fixation with the coordinates in image plane (u, v) in the image plane, a 2D-Gaussian surface $N_{(u,v,\Sigma)}$ is centered with the mean vector $\boldsymbol{\mu} = (u, v)^T$ and a diagonal covariance matrix $\mathfrak{C}$ with equal $\mathfrak{c}_{11} = \mathfrak{c}_{22} = \mathfrak{c}^2$ values. The scale parameter $\mathfrak{c}$ is defined from the geometry of the experiment to represent the projection of the fovea, into the image plane. Summing up and normalizing multi-Gaussian surface from different observers for the same image,

Fig. 12.3. Example of GFDMs on Cat2000 database (with GFDM overlaid).

its GFDM G is obtained. An example of such maps on the dataset from [18] is illustrated in Figure 12.3.

We call this CNN-based fusion operator NET. As a light fusion CNN m, we use a simple architecture. The input tensor consisting of intermediate importance maps S_l is pushed through three successive convolution layers with pooling. They have the total effect of pooling the input tensor by a factor of 4 and multiplying the depth by a factor of 8. Lastly, a weighted sum is computed to output a final predicted 2D map. The loss function is the Euclidean loss, which is the mean square error between GDFM G and the application of the NET operator to the input set of importance maps S_l. The fusion network m is trained on GFDMs of a training set of a given dataset. We present our datasets in Section 12.6.1.

12.5.3 Implementation of FEM and MLFEM on CNN classifiers

Both methods, FEM and MLFEM, are white-box methods and can be applied to any architecture of a CNN in visual classification tasks. Here we present the implementation of them for ResNet50 [21] architecture.

ResNet50 contains 16 residual blocks. FEM is applied on the last convolutional layer of ResNet50 just before the first fully connected layer. Regarding MLFEM, as a block is the natural unit of choice to apply FEM, we apply it to the output of each of them, after the activation function. This gives us 16 different applications of FEM to fuse. The goal is to maximize the different semantic meaning that one can extract: For ResNet50, we take advantage of the structure of the network.

ResNet50 used in our experiments is trained with the Adam optimizer [22], with a binary cross-entropy for binary classification tasks and a categorical cross-entropy loss for multi-label classification tasks.

12.5.4 *Selection of important features for MLFEM*

MLEFEM is a straightforward extension of FEM on the ensemble of selected convolutional layers in a CNN. As FEM is applied on each layer separately, the choice of the threshold for statistical filtering of features in the layers is fulfilled with exactly the same approach as proposed for FEM, see Section 12.3. Obviously, as this choice has to be fulfilled at each layer of the CNN, the related computational cost grows and the optimal choice of the threshold has to be done with regard to the target quality criteria of the method and its computational overload.

12.6 Experimental Protocol

12.6.1 *Datasets*

We have chosen two different datasets to work on. The first one is Salicon [16]. It is composed of 15,000 images of different (up to 80) categories; 10,000 of them are supplied with GFDMs. To construct the GFDMs, the subjects were free to "look around" the image and expressed their attention by mouse pointing. We illustrate the GFMDs from this dataset in Figure 12.4.

Finally, we use Cat2000 [18] as a second dataset. We have illustrated the GFDMs in Figure 12.3. It is composed of 4,000 images, 2,000 with GFDM, each divided into 20 equally populated categories.

Fig. 12.4. Sample of the data in Salicon database (with GFDM overlaid).

The GFDMs were obtained from gaze fixations, and the images are less cluttered.

12.6.2 Evaluation of FEM and MLFEM explanations

12.6.2.1 Methodology

Different methodologies of evaluation of explanation methods have been proposed in the literature [4, 23]. It consists in the comparison of the pixel importance maps obtained by the network sensing with the maps expressing human perception of the same visual content. This perception is expressed by gaze fixation density maps we have presented in Section 12.5.2. Today, this methodology is possible thanks to public databases with the recorded gaze fixation of observers, like [16,18,20], and we apply it to our MLFEM with different fusion operators and compare with different explanation methods which generate pixel importance maps as MLFEM does.

12.6.2.2 Evaluation metrics

Evaluating the relevance of pixel importance explanation maps is an open problem. There are no widely agreed upon metrics for assessing their quality. We propose to employ metrics widely used in psycho-visual community for comparison of saliency maps [24]: the Pearson Correlation Coefficient (PCC) and the similarity metric (SIM). The Pearson Correlation Coefficient (PCC) is defined as

$$\mathrm{corr}\,(A,B) = \frac{\sum_x^W \sum_y^H \left(A(x,y) - \overline{A}\right)\left(B(x,y) - \overline{B}\right)}{\sqrt{\sum_x^W \sum_y^H \left(A(x,y) - \overline{A}\right)^2}\sqrt{\sum_x^W \sum_y^H \left(B(u,v) - \overline{B}\right)^2}}, \tag{12.19}$$

with $A(x,y)$ being the value of the pixel importance at position (x,y) in the saliency map A.

The similarity metric (SIM) is defined as

$$\mathrm{sim}\,(A,B) = \sum_x^W \sum_y^H \min\left(A(x,y), B(x,y)\right). \tag{12.20}$$

12.6.2.3 Design of experiments

To evaluate the proposed methods, we perform three kinds of experiments: (i) overall comparison of FEM and MLFEM with different strategies of threshold choice, (ii) comparison with a popular SOA method — GradCam [7], analysis for correct or wrong classification results, and (iii) sensitivity to clutter in the image. All comparisons are fulfilled with regard to the ground truth, which are GFDMs with the *corr* and *sim* metrics introduced in Eqs. (12.19) and (12.20).

Overall with different strategies of threshold choice: We compare pixel importance maps generated by MLFEM and FEM with different threshold choice strategies.

Comparison with SOA — GradCam: Here we compare our different strategies in FEM and multi-layered FEM with a popular SOA method: GradCam.

Dependence on correct/wrong prediction: We divide the test dataset into images that were correctly categorized by the neural network M and those that were not. We analyze if a drop in correlation with the ground-truth GFDMs is observed.

In the case of a drop, we can deduce that the convolutional part of the network did not pick up on the relevant features of the image. The fully connected part of the network will then not have the correct information in feature space to classify the image.

If the metrics do not change between correctly and wrongly classified images, we can presume that the fully connected part of M, given supposedly correct feature space information, was not able to classify the images. We can then add more fully connected layers to help the network categorize the inputs.

Sensitivity to clutter: In case the image is cluttered, the GFDMs are dispersed as human attention is attracted by multiple singularities/objects in the image. It is reasonable to expect that the CNN allocates more importance to the strongest relevant region in the image. In this case, the similarity metrics between our explanation map and GFDMs will be lower than those for images with low clutter effects.

12.7 Results

This section presents the results of both methods FEM and MLFEM and compares them with the popular method GradCam. First, we present our experiment on the statistical threshold choice according to the method in Section 12.4. Then, we compare the explainers FEM and MLFEM, their versions with statistically adaptive thresholds, and GradCam.

12.7.1 *Statistically adaptive threshold choice*

The first step of the method consists in removing zero-valued maps from the feature tensor at each convolutional block in ResNet50 in MLFEM and at the last convolutional layer in the FEM method, see the algorithm in Section 12.4. The results of dropping out zero-valued maps are presented in Table 12.1 in percentages. We note that the number of zero-valued maps to remove is quite high at the last conv block, which contains 2048 feature maps in the feature tensor in ResNet50. They are 1358.35 ± 112.72 on Salicon dataset and 1261.31 ± 125.93 on Cat2000, respectively. Note that these figures are the same for FEM and MLFEM, which can be seen as restriction of MLFEM for the last convolution block. In the blocks underneath 14th block, there were no zero-valued maps detected in feature tensors. The feature value is compared to zero with a single precision, namely seven significant digits on original feature values without quantization.

At the second step on remaining feature maps, the threshold for selection of important features is computed according to the fitness of

Table 12.1. Percentage of zero-valued feature maps on the blocks of ResNet50, Salicon, and Cat2000 datasets.

Layer	Salicon	Cat2000
16	$66.33 \pm 0.05\%$	$61.58 \pm 6.15\%$
15	$0.16 \pm 0.00\%$	$0.18 \pm 0.19\%$
14	$0.01 \pm 0.00\%$	$0.02 \pm 0.04\%$

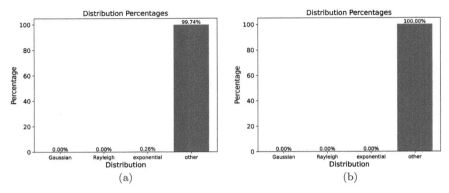

Fig. 12.5. Distribution of percentage of fitness of experimental pdf of features to the theoretical distributions: Gaussian, Rayleigh, exponential and none of them (other): (a) on Salicon [16] dataset; (b) on Cat2000 [18] dataset.

feature experimental law to one of our considered distributions (exponential, Rayleigh, or Gaussian). In the case when none of the distributions is accepted according to K–S test, the threshold is chosen on experimental distribution at 0.975 probability level, see Section 12.4. Nevertheless, such a fit is quite rare. As we illustrated in Figure 12.5 in SALICON dataset (Figure 12.5(a)), only our statistic, that is, feature value, is distributed with exponential distribution according to the K–S test with probability 0.95 (p-value is at least 0.05) in 0.26% of the cases. In Cat2000 dataset (Figure 12.5(b)), our considered distributions according to the K–S test are absent. This means that in all other cases of selected non-zero valued feature maps, the threshold is computed on the experimental cdf. The computation of the statistically adaptive threshold th is much slower than the computation of the threshold according to 2σ-rule. Although calculating an experimental threshold in the case of an undetermined distribution would be a better solution, it takes stronger computational time.

Table 12.2 shows average computational times in seconds for the initial threshold choice according to 2σ-rule and for the statistically adaptive threshold computation on 5 000 images of Salicon test set on the same GPU.

This table illustrates that the new method for the threshold computation is slower, at the deeper blocks, up to 100 times.

The natural question to ask is if the choice of statistically adaptive threshold computed from experimental pdfs improves the quality of

Table 12.2. Average time (in seconds) for computation of two times of thresholds.

Layer	2σ	th	Layer	2σ	th
1	0.001 ± 0.000	0.293 ± 0.044	9	0.002 ± 0.000	0.552 ± 0.010
2	0.001 ± 0.000	0.855 ± 0.013	10	0.004 ± 0.001	0.527 ± 0.025
3	0.001 ± 0.000	0.862 ± 0.013	11	0.004 ± 0.000	0.521 ± 0.022
4	0.002 ± 0.000	0.561 ± 0.012	12	0.004 ± 0.000	0.524 ± 0.022
5	0.002 ± 0.000	0.558 ± 0.011	13	0.004 ± 0.000	0.506 ± 0.022
6	0.002 ± 0.000	0.556 ± 0.010	14	0.011 ± 0.001	0.74 ± 0.073
7	0.002 ± 0.000	0.556 ± 0.011	15	0.008 ± 0.001	0.73 ± 0.078
8	0.002 ± 0.000	0.554 ± 0.010	16	0.009 ± 0.002	0.747 ± 0.085

explanation maps with regard to our considered criteria: Their similarity to GFDMs are expressed in PCC (12.19) and SIM (12.20) metrics. Figure 12.6 illustrates the distribution of metrics' values on Salicon (Figure 12.6(a)) and Cat2000 (Figure 12.6(b)) datasets. It can be seen that using statistically adaptive threshold computed from experimental cdfs slightly improves FEM on Salicon dataset but deteriorates MLFEM on both databases. For Cat2000, both FEM and MLFEM with cumulative threshold perform worse.

The relative difference in means of PCC with different thresholds computation in case of FEM is less than 9.58%. For SIM, it is less than 9.07%. For MLFEM, it is larger, 40.58% and 23.68% for PCC and SIM, respectively. Therefore, it is interesting to test the hypothesis of equality of means of both metrics on our datasets. We have appealed to the unpaired t-test with non-equal variances [25]. Table 12.3 contains the results of the tests on equality of means of metrics. As the value of t-statistic is negative for FEM and p-value is less than the 0.05 level, then the null hypothesis of equality of the means is rejected and the choice of statistically adapted threshold on pdf brings better concordance with GFDMs in terms of both metrics, see lines 1 and 2 on Salicon dataset. On CAT dataset, the null hypothesis is also rejected, the means are different, but the choice of threshold by 2σ-rule ensures better performance than threshold computed on cdf as the t-statistic is positive.

In the case of MLFEM, lines 3 and 4 of Table 12.3, the 2σ-rule threshold is systematically better. The results of the t-test are in concordance with the illustration of performances in Figure 12.6.

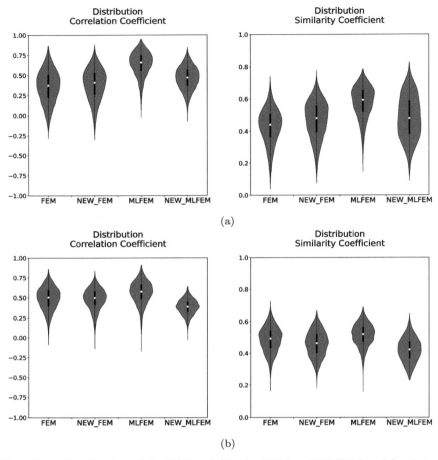

Fig. 12.6. Distribution of the PCC and SIM for FEM and MLFEM and for their NEW versions with statistically adaptive threshold (Section 12.4) on Salicon and Cat2000 datasets.

In the previous experiments on MLFEM in this chapter, we have used fusion of feature importance maps from convolutional blocks by a trained shallow CNN. Indeed, this fusion method was shown to be the best in case of 2σ-rule threshold choice in [6]. We compare now different fusion strategies in MLFEM using our statistically adaptive threshold, with FEM and popular GradCam [7], as the different

Table 12.3. Results of t-test on equality of means of metrics, Salicon and Cat2000 datasets.

Type	Salicon		Cat2000	
	T-statistic	p-value	T-statistic	p-value
FEM-PCC	-6.4827	9.4374e$-$11	1.9876	0.0469
FEM-SIM	-17.9997	2.6186e$-$71	10.6789	2.8778e$-$26
MLFEM-PCC	61.2353	0	48.0993	0
MLFEM-SIM	40.3847	0	40.7827	4.7553e$-$304

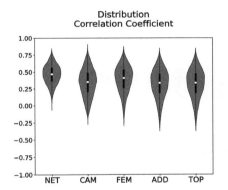 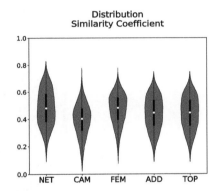

Fig. 12.7. Distribution of the different metrics for each explanation method for the Salicon (with statistically adaptive threshold 12.4) dataset.

choice of the threshold might influence the quality of fusion strategy. The results illustrated in Figure 12.7 for Salicon dataset indeed show that the fusion method "NET" outperforms all other methods even in the case of statistically adaptive threshold, but it is quite close to the FEM. The relative difference of means for PCC metric is of 14.46% with 0.4104 of mean PCC for FEM and 0.4697 for MLFEM. For SIM, the relative difference is even lower: 0.38% (with 0.4805 for FEM and 0.4787 for MLFEM). Figure 12.8 contains an example of explanation maps at each convolutional block of ResNet50 obtained with 2σ-rule-based threshold (upper row) and the statistically adaptive threshold (bottom row). It can be observed that the maps with statistically adaptive threshold highlight more features in the feature maps.

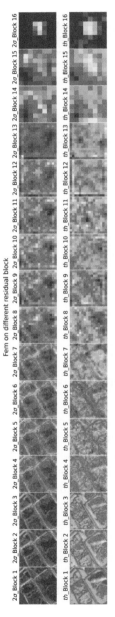

Fig. 12.8. FEM applied on every convolutional block of a typical ResNet50 architecture. (2σ upper row and th bottom row) On an example from Salicon dataset.

12.7.2 *Comparative evaluation of MLFEM*

As we have stated in Section 12.7.1, in all metrics on our two datasets, MLFEM with statistically adaptive threshold yields poorer results. Hence, in this section, we come back to the initial 2σ-rule and compare MLFEM with different fusion strategies: FEM and GradCam. The results are reported according to [6].

12.7.2.1 *Overall method comparison*

We compute two similarity metrics PCC and SIM for each classified image of each dataset for every method (FEM, GradCam noted CAM, and MLFEM with NET, ADD, and TOP variants) with regard to the GFDMs. To compare the methods between them, we compute a 2×2 matrix, comparing the number of times that a method m_a was better at explaining classification than a method m_b in terms of higher value of each metric. This will give us six matrices: two for each dataset, one for the comparison using PCC, and another for the SIM, see Figures 12.9(a) and 12.9(b).

In Figures 12.10(a) and 12.10(b), the distribution of the metrics in Salicon and Cat2000 datasets is plotted for every method compared to the GFDMs.

We can see here in Figures 12.9(a) and 12.10(a) that the proposed MLFEM is better suited for the explanation of ResNet50 on the Salicon dataset. We achieve a mean correlation coefficient of 0.70, whereas the GradCam method only achieves 0.37. We can also see that without resorting to learned fusion operator, FEM and ADD/TOP are better with, respectively, $0.38, 0.43$, and 0.41 values of PCC. The SIM behavior is the same. The trained NET fusion operator is the best in terms of both metrics. For the Cat2000 dataset, the conclusion is the same as for Salicon, as illustrated in 12.9(b) and 12.10(b).

12.7.2.2 *Dependency to correct and wrong predictions*

For Salicon, we are in a multi-label classification task with objects of several classes in the same image. We count the number of times when every class present in an image is detected by the network, i.e., the corresponding output neuron has an activation value larger

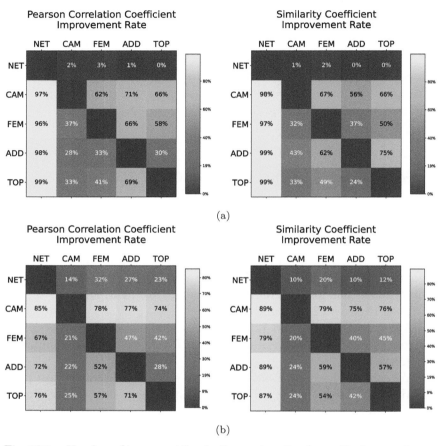

Fig. 12.9. Number of images with a better explanation by method m_a, column, than with method m_b, line, on the Salicon (a) and Cat2000 (b) datasets.

than 0.5, as a correct prediction. When at least one class is not correctly predicted, this image is considered wrongly classified.

We do not see any significant effect of the correct/wrong categorization of the image on the quality of the explanation map (Figures 12.11(a) and 12.11(b) display the average PCC and SIM for both correctly and wrongly classified samples). For Salicon dataset (Figure 12.11(a)), we have a small drop in the quality of explanation for missclassified images (in red), a 3% drop in the PCC and no change in the SIM. For Cat2000, Figure 12.11(b), a small increase in

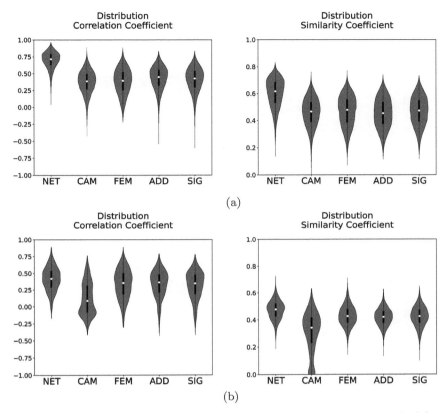

Fig. 12.10. Distribution of the different metrics for each explanation method for the Salicon and Cat2000 datasets.

the quality of explanation for missclassified images is observed. The PCC is 4% better, but there is practically no change for the SIM.

Sensitivity to clutter: This experiment is conducted on Salicon dataset with different classes of objects present in the same image. We divide images into 10 categories, with ith category containing images that have i different classes present in it. Then we can plot the mean of the different metrics as a function of the number of classes appearing in an image.

We can see in Figure 12.12 that our hypothesis holds: the quality of our explanation drops when the number of classes present in

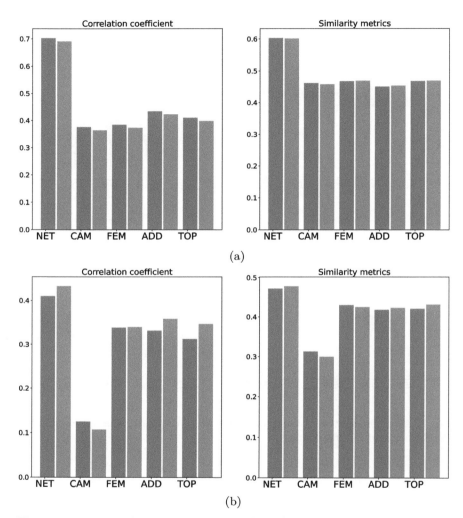

Fig. 12.11. Metrics for correctly classified (green) and wrongly classified images (red) for the Salicon and Cat2000 datasets with NET, CAM, FEM, ADD, and TOP.

an image increases. However, even in high clutter situation, our proposed method with *NET* fusion gives the best scores. The *NET* method only loses 11% of its performance measured by PCC, where the FEM loses about 33%.

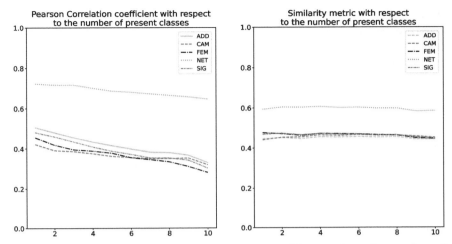

Fig. 12.12. The evolution of similarity metrics with regard to the number of classes present in the image to explain on the SALICON dataset.

The SIM, however, shows a lower dependence with regard to the number of classes present in an input image. The implication of this merits further research in future works.

12.8 Conclusion

In this chapter, we have explored and proposed modifications to the FEM [13] and MLFEM [6] we previously proposed; they are both based on the selection of strong features in the convolutional layers of the CNN. We have proposed a new strategy for the choice of the threshold for filtering features. While in the original FEM and MLFEM this choice was based on the underlying assumption of Gaussian distribution of features in the feature maps, we proposed to test fitness of the experimental feature law to some of known parametric distributions. We have considered the distributions with shapes which are close to our experimental pdf as it contains strong pick in zero-valued bin.

Statistical test of Kolmogorov–Smirnov has shown, on two databases, that the fit to the distribution of known shape is very rare, only in 0.26% of cases on Salicon dataset. Therefore, we have proposed an adaptive algorithm of threshold choice which comprises the K–S test and allows for threshold computation either from parametric cdf or from experimental pdf. The latter happens when the experimental pdf does not fit to any of the proposed parametric laws.

As the computation of the threshold on experimental pdf happens in the majority of cases and is very much time-consuming, we have proposed dropping of zero-valued feature maps before statistical threshold computation. Here interesting results have been obtained. Indeed, on the last convolutional layer of ResNet50, around 60% of feature maps can be dropped before the feature tensor is submitted to the classification end.

The strategy of using adaptive statistical threshold was evaluated in terms of two metrics: PCC and SIM. Using an adaptive threshold leads to better results in terms of mean values of our metrics, compared to a threshold chosen according to 2σ-rule, for the FEM on the Salicon dataset. In the case of the Cat2000 dataset, the adaptive threshold shows a drop in explanation quality. There is also a drop when using the MLFEM algorithm on both datasets. In the first case, this may be due to the sensitivity to the input data, which contains parasite features induced by format conversion artifacts on this dataset. In the second case, a possible problem could be the principle of the MLFEM algorithm itself, which may cumulate unwanted features from upper layers of the network using the adaptive threshold. Both of these problems require further investigation.

Thus, the use of an adaptive threshold for FEM would be preferable in the absence of time constraints for the algorithm. Otherwise, a threshold based on the 2σ-rule is a qualitative alternative.

As MLFEM is a multi-layer alternative to FEM, collecting strong features from different convolutional layers of the CNN classifier, we have also studied the influence of the choice of the threshold method on its performances with different fusion strategies which were proposed for feature collecting.

The performance of both methods was assessed on ResNet as the latter is nowadays the most efficient CNN classifier. Nevertheless, the method remains generic and applicable to any CNN whether it is residual or not. We have also benchmarked our methods with regard to the popular benchmark, such as GradCam, and showed their better performance in terms of comparison with gaze fixation density maps.

In the future works, it may be interesting to apply and adapt MLFEM to transformer networks and to use it with another kind of data than images. In such a case, we will face the problem of definition of the ground truth for the evaluation methodology proposed. Hence, the proposed method opens multiple research questions which have to be addressed in the future.

Acknowledgment

We thank Dr. Damien Garreau for fruitful discussions on statistical properties of features in CNNs.

References

[1] A. Aliper, S. Plis, A. Artemov, A. Ulloa, P. Mamoshina and A. Zhavoronkov, Deep learning applications for predicting pharmacological properties of drugs and drug repurposing using transcriptomic data, *Molecular Pharmaceutics* **13**(7), pp. 2524–2530 (2016), https://doi.org/10.1021/acs.molpharmaceut.6b00248.

[2] Y. Goldberg, Neural network methods for natural language processing, *Synthesis Lectures on Human Language Technologies* **10**(1), pp. 1–309 (2017), https://doi.org/10.1007/978-3-031-02165-7.

[3] J. Benois-Pineau, R. Bourqui, D. Petkovic and G. Quénot (eds.), *Explainable Deep Learning AI*. Elsevier; Academic Press (2023), https://doi.org/10.1016/C2021-0-01538-8.

[4] M. P. Ayyar, J. Benois-Pineau and A. Zemmari, Review of white box methods for explanations of convolutional neural networks in image classification tasks, *Journal of Electronic Imaging* **30**(5), pp. 050901–050901 (2021), https://doi.org/10.1117/1.JEI.30.5.050901.

[5] M. T. Ribeiro, S. Singh and C. Guestrin, "Why should I trust you?": Explaining the predictions of any classifier, in *Proceedings of the 22nd ACM SIGKDD International Conference on Knowledge Discovery and Data Mining, KDD'16*. Association for Computing Machinery, New York, p. 1135–1144 (2016), https://doi.org/10.1145/2939672.2939778.

[6] L. Bourroux, J. Benois-Pineau, R. Bourqui and R. Giot, Multi layered feature explanation method for convolutional neural networks ⋆, in *International Conference on Pattern Recognition and Artificial Intelligence (ICPRAI)*, Paris, France (2022), https://doi.org/10.1007/978-3-031-09037-0_49.

[7] R. R. Selvaraju, M. Cogswell, A. Das, R. Vedantam, D. Parikh and D. Batra, Grad-cam: Visual explanations from deep networks via gradient-based localization, *International Journal of Computer Vision* **128**(2), pp. 336–359 (2019), http://dx.doi.org/10.1007/s11263-019-01228-7.

[8] L.-E. Pommé, R. Bourqui and R. Giot, H^2O: Heatmap by hierarchical occlusion, in *Proceedings of the 20th International Conference on Content-based Multimedia Indexing (CBMI2023)*, Orléans, France. ACM, p. 6 (2023), https://hal.science/hal-04212098.

[9] M. D. Zeiler and R. Fergus, Visualizing and understanding convolutional networks, in D. Fleet, T. Pajdla, B. Schiele and T. Tuytelaars (eds.), *Computer Vision — ECCV 2014*. Springer International Publishing, pp. 818–833 (2014), https://doi.org/10.1007/978-3-319-10590-1_53.

[10] D. Smilkov, N. Thorat, B. Kim, F. B. Viégas and M. Wattenberg, Smoothgrad: Removing noise by adding noise pp. 1–10 (2017), *CoRR*, http://arxiv.org/abs/1706.03825.

[11] M. Sundararajan, A. Taly and Q. Yan, Axiomatic attribution for deep networks, in *Proceedings of the 34th International Conference on Machine Learning, ICML'17*, Vol. 70. JMLR.org, pp. 3319–3328 (2017), https://doi.org/10.48550/arXiv.1703.01365.

[12] S. Bach, A. Binder, G. Montavon, F. Klauschen, K.-R. Müller and W. Samek, On pixel-wise explanations for non-linear classifier decisions by layer-wise relevance propagation, *PLOS One* **10**(7), pp. 1–46 (2015), https://doi.org/10.1371/journal.pone.0130140.

[13] K. Ahmed Asif Fuad, P.-E. Martin, R. Giot, R. Bourqui, J. Benois-Pineau and A. Zemmari, Features understanding in 3d cnns for actions recognition in video, in *2020 10th International Conference on Image Processing Theory, Tools and Applications (IPTA)*, pp. 1–6 (2020), https://doi.org/10.1109/IPTA50016.2020.9286629.

[14] A. Zemmari and J. Benois-Pineau, *Deep Learning in Mining of Visual Content*, 1 edn. Springer, Cham (2020), https://doi.org/10.1007/978-3-030-34376-7.

[15] R. Hogg, J. McKean and A. Craig, *Introduction to Mathematical Statistics*. What's New in Statistics Series. Pearson (2019).

[16] M. Jiang, S. Huang, J. Duan and Q. Zhao, Salicon: Saliency in context, in *2015 IEEE Conference on Computer Vision and Pattern Recognition (CVPR)*, Boston, MA, USA, pp. 1072–1080 (2015), https://doi.org/10.1109/ CVPR.2015.7298710.

[17] N. Smirnov, Table for estimating the goodness of fit of empirical distributions, *The Annals of Mathematical Statistics* **19**(2), pp. 279–281 (1948), https://doi.org/10.1214/aoms/1177730256.

[18] A. Borji and L. Itti, CAT2000: A large scale fixation dataset for boosting saliency research (2015) *CoRR* http://arxiv.org/abs/1505.03581.

[19] K. Aderghal, K. Afdel, J. Benois-Pineau, G. Catheline, A. D. N. Initiative *et al.*, Improving Alzheimer's stage categorization with convolutional neural network using transfer learning and different magnetic resonance imaging modalities, *Heliyon* **6**(12), p. e05652 (2020), https://doi.org/10.1016/j.heliyon.2020.e05652.

[20] A. M. Obeso, J. Benois-Pineau, M. S. García Vázquez and A. Álvaro Ramírez Acosta, Visual vs internal attention mechanisms in deep neural networks for image classification and object detection, *Pattern Recognition* **123**, p. 108411 (2022), https://doi.org/10.1016/j.patcog.2021.108411.

[21] F. Rousseau, L. Drumetz and R. Fablet, Residual networks as flows of diffeomorphisms, *Journal of Mathematical Imaging and Vision* **62**, pp. 365–375 (2020), https://doi.org/10.1007/s10851-019-00890-3.

[22] D. Kingma and J. Ba, Adam: A method for stochastic optimization (2014), https://doi.org/10.48550/arXiv.1412.6980.

[23] G. Jouis, H. Mouchère, F. Picarougne and A. Hardouin, Anchors vs attention: Comparing xai on a real-life use case, in *Pattern Recognition. ICPR International Workshops and Challenges: Virtual Event*, January 10–15, 2021, Proceedings, Part III. Springer-Verlag, Berlin, p. 219–227 (2021), https://doi.org/10.1007/978-3-030-68796-0_16.

[24] O. L. Meur and T. Baccino, Methods for comparing scanpaths and saliency maps: Strengths and weaknesses, *Behavior Research Methods*, pp. 251–266 (2012), https://doi.org/10.3758/s13428-012-0226-9.

[25] K. K. Yuen, The two-sample trimmed t for unequal population variances, *Biometrika* **61**(1), pp. 165–170 (1974), https://doi.org/10.2307/2334299.

Milton Keynes UK
Ingram Content Group UK Ltd.
UKHW021844090924
448121UK00002B/3